Differential Equations with Mathematica®

Third Edition

Updated for Mathematica 6

Brian R. Hunt
Ronald L. Lipsman
John E. Osborn
Donald A. Outing
Jonathan M. Rosenberg

with

Kevin R. Coombes
Garrett J. Stuck

John Wiley & Sons, Inc.

PUBLISHER	Laurie Rosatone
ASSOCIATE EDITOR	Shannon Corliss
MARKETING MANAGER	Jaclyn Elkins
SENIOR PRODUCTION EDITOR	Ken Santor
COVER DESIGNER	Kevin Murphy

This book was set in LaTeX by the authors and printed and bound by Malloy Inc.

This book is printed on acid free paper. ∞

Mathematica® is a registered trademark of Wolfram Research, Inc.
The computations in this book were performed using Mathematica 6.

ISBN 978-0-471-77316-0

Printed in the United States of America

10 9 8 7 6 5 4 3 2 1

Preface

As the subject matter of differential equations continues to grow, as new technologies become commonplace, as old areas of application are expanded, and as new ones appear on the horizon, the content and viewpoint of courses and their textbooks must also evolve.

William E. Boyce and Richard C. DiPrima, *Elementary Differential Equations*, Ninth Edition.

Traditional introductory courses in ordinary differential equations (ODE) have concentrated on teaching a repertoire of techniques for finding formula solutions of various classes of differential equations. Typically, the result was rote application of formula techniques without a serious qualitative understanding of such fundamental aspects of the subject as stability, asymptotics, dependence on parameters, and numerical methods. These fundamental ideas are difficult to teach because they have a great deal of geometrical content and, especially in the case of numerical methods, involve a great deal of computation. Modern mathematical software systems, which are particularly effective for geometrical and numerical analysis, can help to overcome these difficulties. This book changes the emphasis in the traditional ODE course by using a mathematical software system to introduce numerical methods, geometric interpretation, symbolic computation, and qualitative analysis into the course in a basic way.

The mathematical software system we use is *Mathematica*®. (This book is also available in MATLAB and Maple versions.) We assume that the user has no prior experience with *Mathematica*. We include concise instructions for using *Mathematica* on any standard computer platform: Windows, Macintosh, or UNIX, including Linux. This book is not a comprehensive introduction or reference manual to either *Mathematica* or any of the computer platforms. Instead, it focuses on the specific features of *Mathematica* that are useful for analyzing differential equations. In this third edition we discuss some of the features of *Mathematica* 6 that make it even easier than it was previously to use *Mathematica* to produce integrated documents that incorporate text, mathematical calculations, and graphics.

This supplement can easily be used in conjunction with most ODE textbooks. It addresses the standard topics in ordinary differential equations, but with a substantially different emphasis.

We had two basic goals in mind when we introduced this supplement into our course. First, we wanted to deepen students' understanding of differential equations by giving them a new tool, a mathematical software system, for analyzing differential equations. Second, we wanted to bring students to a level of expertise in the mathematical software system that would allow them to use it in other mathematics, engineering, or science courses. We believe that we have achieved these goals in our own classes. We hope this supplement will be useful to students and instructors on other campuses in achieving the same goals.

Acknowledgment and Disclaimer. We are pleased to acknowledge support of our research by the National Science Foundation, which contributed over many years to the writing of this book. Our work on the third edition was partially supported by NSF Grants DMS-0616585, DMS-0611094, and DMS-0805003. Any opinions, findings, conclusions, or recommendations expressed in this material are those of the authors and do not necessarily reflect the views of the National Science Foundation.

Brian R. Hunt
Ronald L. Lipsman
John E. Osborn
Donald A. Outing
Jonathan M. Rosenberg

College Park, Maryland and West Point, New York
August 2008

Contents

Chapter 1

Introduction

We begin by describing the philosophy behind our approach to the study of ordinary differential equations. This philosophy has its roots in the way we understand and apply differential equations; it has influenced our teaching and guided the development of this book. This chapter also contains two user's guides, one for students and one for instructors.

1.1 Guiding Philosophy

In scientific inquiry, when we are interested in understanding, describing, or predicting some complex phenomenon, we use the technique of mathematical modeling. In this approach, we describe the state of a physical, biological, economic, or other system by one or more functions of one or several variables. For example:

- The position $s(t)$ of a particle is a function of the time t.

- The temperature $T(x, y, z, t)$ in a body is a function of the position (x, y, z) and the time t.

- The populations $x(t)$ and $y(t)$ of two competing species are functions of time.

- The gravitational or electromagnetic force on an object is a function of its position.

- The money supply is a function of time.

Next, we attempt to formulate, in mathematical terms, the fundamental law governing the phenomenon. Typically, this formulation results in one or more differential equations; *i.e.*, equations involving derivatives of the functions describing the state of the system with respect to the variables they depend on. Frequently, the functions depend on only one variable, and the differential equation is called *ordinary*. To be specific, if $x(t)$ denotes the function describing the state of our system, an *ordinary differential equation* for $x(t)$ might involve $x'(t)$, $x''(t)$, higher derivatives, or other known functions of t. By contrast, in a *partial differential equation*, the functions depend on several variables.

1

In addition to the fundamental law, we usually describe the initial state of the system. We express this state mathematically by specifying *initial values* $x(0)$, $x'(0)$, *etc*. In this way, we arrive at an *initial value problem*—an ordinary differential equation together with initial conditions. If we can solve the initial value problem, then the solution is a function $x(t)$ that predicts the future state of the system. Using the solution, we can describe qualitative or quantitative properties of the system. At this stage, we compare the values predicted by $x(t)$ against experimental data accumulated by observing the system. If the experimental data and the function values match, we assert that our model accurately models the system. If they do not match, we go back, refine the model, and start again. Even if we are satisfied with the model, new technology, new requirements, or newly discovered features of the system may render our old model obsolete. Again, we respond by reexamining the model.

The subject of differential equations consists in large part of building, solving, and analyzing mathematical models. Many results and methods have been developed for this purpose. These results and methods fall within one or more of the following themes:

1. Existence and uniqueness of solutions,

2. Dependence of solutions on initial values,

3. Derivation of formulas for solutions,

4. Numerical calculation of solutions,

5. Graphical analysis of solutions,

6. Qualitative analysis of differential equations and their solutions.

The basic results on existence and uniqueness (theme 1) and dependence on initial values (theme 2) form the foundation of the subject of differential equations. The eminent French mathematician Jacques Hadamard referred to a problem as *well-posed* if good results are available concerning these two themes, *i.e.*, if one has existence and uniqueness of solutions and if small changes in initial values make only a small change in the solution. The derivation of formula solutions (theme 3) is a rich and important part of the subject; a variety of methods have been developed for finding formula solutions to special classes of equations. Although many equations can be solved exactly, many others cannot. However, any equation, solvable or not, can be analyzed using numerical methods (theme 4), graphical methods (theme 5), and qualitative methods (theme 6). The ability to obtain qualitative and quantitative information without the aid of an explicit formula solution is crucial. That information may suffice to analyze and describe the original phenomenon (which led to the model, which gave rise to the differential equation).

Traditionally, introductory courses in differential equations focused on methods for deriving exact solutions to special types of equations and included some simple numerical and qualitative methods. The human limitations involved in compiling numerical or graphical data were formidable obstacles to implementing more advanced qualitative or quantitative methods. Computer platforms have reduced these obstacles. Sophisticated software and mainframe computers enhanced the use of quantitative and qualitative methods in the theory and applications of differential equations. With the arrival of comprehensive mathematical software systems on personal computers, this modern approach has become accessible to students.

In this book, we use the mathematical software system *Mathematica* to implement this approach. We use *Mathematica*'s symbolic, numerical, and graphical capabilities to analyze differential equations and their solutions.

Finally, engineers and scientists have to develop not only skills in analyzing problems and interpreting solutions, but also the ability to present coherent conclusions in a logical and convincing style. Students should learn how to submit solutions to the computer assignments in such a style. This is excellent preparation for professional requirements that lie ahead.

1.2   Student's Guide

The chapters of this book can be divided into three classes: general discussion of *Mathematica*, supplementary material on ordinary differential equations (ODE), and computer problem sets. Here is a brief description of the contents.

Chapter 2 explains how to start and run *Mathematica* on your computer. Chapter 3 introduces basic *Mathematica* commands. Unless you have previous experience with *Mathematica*, you should work through Chapter 3 while sitting at your computer. Then you should read Chapter 4, which contains detailed instructions for using *Mathematica* Notebooks and printing or presenting your work. After that, work the problems in Problem Set A to practice the skills you've learned in Chapters 3 and 4. These steps will bring you to a basic level of competence in the use of *Mathematica*, sufficient for the first three of the differential equations chapters, Chapters 5–7, and for Problem Set B. Some more advanced aspects of *Mathematica*, needed for some of the problems starting in Problem Set C, are discussed in Chapter 8.

Since the primary purpose of this book is to study differential equations, we have not attempted to describe all the major aspects of *Mathematica*. You can explore *Mathematica* in more depth using its demos, tutorials, and online help, or by consulting more comprehensive books such as the *Mathematica* 6 manuals available at

http://reference.wolfram.com/mathematica/guide/Mathematica.html

The eight ODE chapters (5–7, 9–13) are intended to supplement the material in your text. The emphasis in this book differs from that found in a traditional ODE text. The main difference is that less emphasis is placed on the search for exact formula solutions, and greater emphasis is placed on qualitative, graphical, and numerical analysis of the equations and their solutions. Furthermore, the commands for analyzing differential equations with *Mathematica* appear in these chapters.

The six computer problem sets form an integral part of the book. Solving these problems will expose you to the qualitative, graphical, and numerical features of the subject. Each set contains about twenty problems.

You can most profitably attack the problem sets if you plan to do them in two sessions. Begin by reading the problems and thinking about the issues involved. Then go to the computer and start solving the problems. If you get stuck, save your work and go on to the next problem. If there are things you'd like to discuss with your instructor, print out the relevant parts of your input and output. Talk to your instructor or your peers about anything you don't understand. This first session should be attempted well before the assignment is due. After you have reviewed your output and obtained answers to your questions, you are ready for your second session. At this point, you should fill the gaps, correct your mistakes, and polish your Notebook. Although you may find yourself spending extra time on the first few problems, if you read the *Mathematica* chapters carefully, and follow the suggestions above, you should steadily increase your level of competence in using *Mathematica*.

The end of this book contains two useful sections: a Glossary and a collection of Sample Solutions. The Glossary contains a brief summary of relevant *Mathematica* commands, built-in functions, and programming constructs. The Sample Solutions show how we solved several problems from this book. These samples can serve as guides when you prepare your own solutions. Emulate them. Strive to prepare coherent, organized solutions. Combine *Mathematica*'s input, output, and graphics with your own textual commentary and analysis of the problem. Edit the final version of your solution to remove syntax errors and false starts. You will soon take pride in submitting complete, polished solutions to the problems.

1.3   Instructor's Guide

The philosophy that guided the writing of this book is explained at the beginning of this chapter. Here is a capsule summary of that philosophy. We seek:

- To guide students into a more interpretive mode of thinking.

- To use a mathematical software package to enhance students' ability to compute symbolic and numerical solutions, and to perform qualitative and graphical analysis of differential equations.

- To develop course material that reflects the current state of ODE and emphasizes the mathematical modeling of physical problems.

- To minimize the time required to learn to use the software package.

As mentioned in Section 1.2, our material consists of *Mathematica* discussion, ODE supplements, and computer problem sets. Here are our recommendations for integrating this material into a typical first course in differential equations for scientists or engineers.

1.3.1   *Mathematica*

Our students read Chapters 2, 3, and 4, and work Problem Set A within the first week of the semester. Although Chapter 8 is not essential for Problem Set B, students often find it useful. Attention to these chapters quickly leads students to a basic level of proficiency.

1.3.2 ODE Chapters

These eight chapters (5–7 and 9–13) supplement the material in a traditional text. We use *Mathematica* to study differential equations using symbolic, numerical, graphical, and qualitative methods. We emphasize the following topics: direction fields, stability, numerical methods, comparison methods, and phase portraits. These topics are not emphasized to the same degree in traditional texts. We incorporate this new emphasis into our class discussions, devoting some class time to each chapter. Specific guidelines are difficult to prescribe, and the required time varies with each chapter, but on average we spend up to an hour per chapter in class discussion.

The structure of this book requires that numerical methods be discussed early in the course, immediately after the discussion of first order equations. The discussion of numerical methods is directed toward the use of **NDSolve**, *Mathematica*'s primary numerical ODE solver.

1.3.3 Computer Problem Sets

There are six computer problem sets. The topics addressed in the problem sets are:

(A) Practice with *Mathematica*,

(B) First Order Equations,

(C) Numerical Solutions,

(D) Second Order Equations,

(E) Series Solutions and Laplace Transforms,

(F) Systems of Differential Equations.

Problem Set A is a practice set designed to acquaint students with the basic symbolic and graphical capabilities of *Mathematica*, and to reacclimate them to calculus. We assign all problems in Problem Set A, and have it turned in rather quickly. We generally assign 3–5 problems from each of the remaining problem sets.

In addition to analyzing problems critically, it is important that students present their analyses in coherent English and mathematics, displayed appropriately on their printouts. To accomplish this, the Notebook mode of *Mathematica*'s interface is ideal. Chapter 4 contains detailed instructions on how to use it for this purpose. Engineers and scientists do not just solve problems; they must also present their ideas in a cogent and convincing fashion. We expect students to do the same in our course. To encourage students to submit high-quality solutions to the homework problems, we have provided *Sample Solutions* to selected problems at the end of this book. These solutions were prepared in *Mathematica* Notebooks and then reformatted for printing.

We should mention that we use the text *Elementary Differential Equations*, Ninth Edition, by William E. Boyce and Richard C. DiPrima, John Wiley & Sons, Inc., 2008, in our course. The references to Boyce and DiPrima in our book are all to this edition. We have found that our book is easily integrated with this text. If you are still using the Eighth Edition of Boyce and DiPrima, it should be easy to locate the corresponding sections and problems. We believe our book can likewise be conveniently integrated with any other text for a first course in differential equations for scientists or engineers. Some suggestions on how to accomplish this may be found on our web site:

```
http://www.math.umd.edu/undergraduate/schol/ode
```

1.4   A Word about Software Versions

New versions of software appear frequently. When a complex program like *Mathematica* changes, many commands work better than they did before, some work differently, and a few may no longer work at all. The instructions in this book were written for *Mathematica* 6, but most of them will work with other versions of *Mathematica* as well.

Chapter 2

Getting Started with *Mathematica*

In this chapter, we will introduce the tools you need in order to begin using *Mathematica* effectively. These include: some relevant information on computer platforms and software; installation protocols; how to launch *Mathematica*, enter commands, use online help, recover from hang-ups, and finally, how to exit the program. We know you are eager to get started using *Mathematica*, so we will keep this chapter brief. After you complete it, you can go immediately to Chapter 3 to find concrete and simple instructions for using *Mathematica* to do mathematics. We describe the *Mathematica* interface more elaborately in Chapter 4, and we start in earnest on differential equations in Chapter 5.

2.1   Platforms and Versions

It is likely that you will run *Mathematica* on a PC (running Windows or Linux), or on some form of UNIX operating system. If you are running a Macintosh platform, you should find that our instructions for Windows platforms will suffice for your needs. Recent versions of *Mathematica* look virtually identical on Windows and Unix platforms; for definitiveness, we shall assume the reader is using a PC in a Windows environment. This book is written to be compatible with the current version of *Mathematica*, namely *Mathematica* 6. We note that for *Mathematica* 6, there is no difference in functionality between the Student and Professional versions of *Mathematica*.

Remark 2.1 We use the word Windows to refer to all flavors of the Windows operating system. *Mathematica* 6 will run on Windows Vista, Windows XP, Windows Server 2003, Windows Compute Cluster Server 2003, Windows 2000, and Windows Me.

2.2   Installation

If you intend to run *Mathematica* on a PC, especially the Student Version, it is quite possible that you will have to install it yourself. You can easily accomplish this using the product CDs. Follow the installation instructions as you would with any new software installation.

2.3   Starting *Mathematica*

You start *Mathematica* as you would any other software application. On a PC you access it via the **Start** menu or an appropriate icon on your desktop. On a UNIX machine, generally you need only type `mathematica` in a terminal window, although you need to first ensure that *MathLM* (the *Mathematica* license manager) is running. Or you may have an icon or a special button on your desktop that achieves the task.

However you start *Mathematica*, you will briefly see a window that displays the *Mathematica* logo as well as some *Mathematica* product information, and then some new windows will appear on the screen. Figure 2.1 shows a standard *Mathematica* display. The narrow window at the top is the *Mathematica menu bar*. The main blank window is a *Mathematica Notebook*, whose name is "Untitled-1". The smaller window on the right is the **BasicMathInput** palette, which can be used to insert basic mathematical symbols into the Notebook. (If you do not see the palette on your screen, click on **Palettes ▶ Basic-MathInput** and it will appear.)

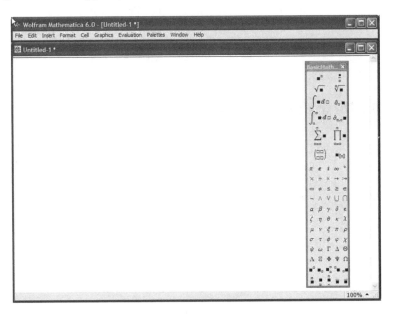

Figure 2.1: A *Mathematica* Display

If this is your first time starting *Mathematica*, a **Startup** palette opens offering **First Five Minutes with Mathematica**, **Documentation Center**, **What's New in 6**, and **Demonstrations Project**. We recommend that you select **First Five Minutes with Mathematica**. This brief tour will give you a chance to explore some basic commands in *Mathematica*.

2.4   *Mathematica* Input

If an "Untitled" Notebook does not appear on your screen, click on **File ▶ New ▶ Notebook (.nb)** to start a new Notebook. Alternatively, you may press CTRL+N. Click in the Notebook Window to make it active. When a window becomes active, its titlebar darkens. Now you can begin entering commands. Try typing **2+2**, then press SHIFT+ENTER to evaluate the input. Alternatively, if your keyboard has a numeric keypad, you may simply press the ENTER key on the numeric keypad to evaluate the input. Next, try typing **FactorInteger[987654321]**. And finally **Cos[100]//N**. Your *Mathematica* display should look like Figure 2.2.

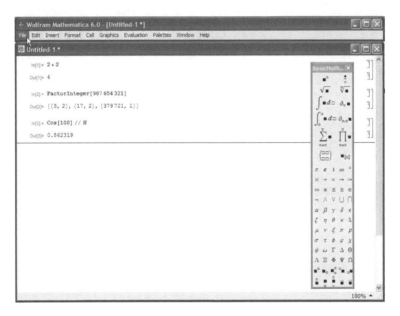

Figure 2.2: The *Mathematica* Desktop with Several Commands Evaluated

Many mathematical notations that cannot easily (if at all) be typed from your keyboard are available in palettes. To open a palette, click on **Palettes** in the menu bar. Select the palette you need from the drop-down window. Then click on a symbol in the palette and it will appear in your *Mathematica* Notebook. For example, you can input a square root or integral using the **BasicMathInput** palette. To close a palette, simply click on the special symbol that closes your windows (usually an × in the upper right-hand corner).

2.5   Online Help

Mathematica has extensive online help. In fact, using only this book and the online help, you should be able to become quite proficient with *Mathematica*.

You can access the online help in one of several ways. For general help, click on **Help ▶ Documentation Center**. This will open the full *Mathematica* documentation. When the Documentation Center opens, you will see a window with several categories and a search box at the top; see Figure 2.3. To familiarize yourself with the system, you can browse the different categories, and look at examples, tutorials, and links to demonstrations.

Figure 2.3: The *Mathematica* Documentation Center

Often, you simply want help on a specific function or topic. Brief help can be found by typing **?** followed by the function name at the *Mathematica* input prompt. For example, you can obtain information about **Solve** by typing **?Solve** followed by SHIFT+ENTER. At the end of the output you will see the symbol ≫, which you can click to take you to more extensive help in the Documentation Center. You can search for functions by including the wildcard variable ***** and then click on an item in the list for more about that function. For example, type **?*Plot** and it will give you every function ending in **Plot**. Try **?*Plus*** to see every function with **Plus** in it somewhere. You can get a similar result in the Documentation Center by typing **plus** in the search box near the top of the window. (As you will learn later, *all Mathematica* commands begin with a capital letter. However, in the Documentation Center, you may type words in lower case. Try typing first **Plus**, then **plus**, in the search box to see the difference in the way that *Mathematica* reacts.) You can also get Documentation Center help for a function by highlighting its name in a Notebook and typing F1.

Remark 2.2 Some functions have several options. To see what is available, evaluate **Options** [*function*], inserting the name of the function you are interested in. For example, **Options** [**Graphics**] shows all of the options for **Graphics** along with their default values.

2.6   Ending a Session

The standard way to conclude a *Mathematica* session is to use the **Exit** option from the **File** menu. You can also click on the special symbol that closes your windows (usually an × in the upper right-hand corner). Before you exit *Mathematica*, you should be sure to save your work, print any graphics or other files you need, and in general clean up after yourself. Some strategies for doing so are addressed in Chapter 4.

Chapter 3

Doing Mathematics with *Mathematica*

This chapter describes some of the basic *Mathematica* commands that you will use in this book. We recommend that you start *Mathematica* and enter the commands displayed in this chapter as you read it.

3.1 Arithmetic

Mathematica can do arithmetic like an electronic calculator. You can add with +, subtract with -, multiply with * (a space also denotes multiplication), divide with /, and exponentiate with ^. To evaluate an arithmetic expression, type the expression and then press SHIFT+ENTER or the ENTER key on the numeric keypad. For example:

```
In[1]:= 2^3 - (4 + 5)/6 + 7*8
```

$$\text{Out}[1] = \frac{125}{2}$$

Notice that *Mathematica* gives the answer in the form of a rational number.

Mathematica differs from a calculator in that it treats numbers *exactly* rather than using decimal approximations. For example, if you enter **4/7**, *Mathematica* will simply output the same fraction. To force *Mathematica* to give a decimal answer, type **N[4/7]**, or equivalently **4/7 //N**, which produces a 6-digit decimal approximation to 4/7. The **N** stands for "numerical". Here are some further examples of exact, or symbolic, arithmetic versus approximate decimal, or numerical, arithmetic.

Mathematica can give the exact value for 2^{100}, even though it has 31 digits, or give a numerical approximation in floating-point notation:

```
In[2]:= 2^100
```

```
Out[2] = 1267650600228229401496703205376
```

```
In[3]:= N[2^100]
```

Out[3]= 1.26765×10^{30}

Next, suppose you want the square root of the fraction $\frac{125}{2}$:

In[4]:= **Sqrt[125/2]**

Out[4]= $5\sqrt{\dfrac{5}{2}}$

Mathematica simplifies the answer symbolically, while keeping it exact. If you want a decimal approximation instead, you can add **//N** to the end of the input line:

In[5]:= **Sqrt[125/2] // N**

Out[5]= 7.90569

When you use **//N** to get a numerical result, *Mathematica* gives you an answer that is accurate to 6 significant figures. You can tell *Mathematica* how many significant figures to give in a particular output as follows. To get a 20-digit decimal approximation to the square root above, type

In[6]:= **N[Sqrt[125/2],20]**

Out[6]= 7.9056941504209483300

Another way to force *Mathematica* to give an approximate numerical value for an expression is to type the expression with an explicit decimal point. For example, when you type an integer like **2**, *Mathematica* assumes that it is exact. If you type **2.0**, *Mathematica* assumes that it is accurate to a fixed number of decimal places. When any number in an arithmetic expression is given with an explicit decimal point, *Mathematica* produces an approximate numerical result for the whole expression.

3.2   Recovering from Problems

Inevitably, when using any mathematical software system, you are bound to encounter minor glitches. Even while entering simple arithmetic commands, you may accidentally mistype an entry or inadvertently violate a *Mathematica* rule. In this brief section, we discuss two methods for coping with these kinds of problems.

3.2.1   Errors in Input

If you make an error in an input line, *Mathematica* will highlight the error and print an error message. For example, here's what happens when you try to evaluate **Sqrt[4, 5]**:

In[7]:= **Sqrt[4, 5]**

        Sqrt::argx : Sqrt called with 2

          arguments; 1 argument is expected. ≫

Out[7]= Sqrt[4, 5]

The square root function should have only one argument. *Mathematica* prints a message to warn you that you have given two arguments. At the end of the output you will see the symbol ≫, which you can click to take you to the Documentation Center. Note that *Mathematica* highlights in red where it thinks the error might be. Improper arguments, missing multiplication operators, brackets, and parentheses are the most common errors.

You can edit an input line by using the mouse to position the cursor at the desired insertion point, typing over the error (or using the "delete" or "backspace" keys), and finally pressing SHIFT+ENTER to reenter the line. You can also use the arrow keys to move around in the Notebook.

3.2.2   Aborting Calculations

If *Mathematica* gets hung up in a calculation, or seems to be taking too long to perform an operation, you can usually abort it by typing ALT+., that is, hold down the ALT key and then press the period key. You can get the same effect by selecting **Abort Evaluation** from the **Evaluation** menu. When doing some operations, it may take *Mathematica* some time to abort the calculation. If aborting the calculation does not work, that is, if the calculation keeps running for more than a few seconds, you can choose **Evaluation ▶ Quit Kernel ▶ Local**. If you do this, *Mathematica* will clear its memory of all the commands that you have run during your current session (though they will remain visible in your Notebook). It will also restart the numbering of your input and output lines.

3.3   Symbolic Computation

Mathematica can carry out a variety of symbolic calculations such as factoring polynomials or solving algebraic equations. Consider the following series of examples, in which the symbol **%** refers to the output of the previous command (see Section 3.3.3):

In[8]:= **(x + y) (x + y)^2**

Out[8]= $(x + y)^3$

In[9]:= **Expand[%]**

Out[9]= $x^3 + 3\ x^2\ y + 3\ x\ y^2 + y^3$

In[10]:= **Factor[%]**

Out[10]= $(x + y)^3$

Notice in the first example that the "*" for multiplication was left out. You can always use a space instead of "*", even between two numbers: **2 5** would evaluate to 10 just like **2*5**. In some cases, such as between parentheses, or in the case of a number times a variable, *Mathematica* will automatically insert the space.

Mathematica often makes minor simplifications to the expressions you type, but does not make any major changes unless you tell it to. You can use *Mathematica*'s **Simplify** command to try to simplify an expression. For example:

In[11] := **Simplify[(x^2 - y^2)/(x - y)]**

Out[11] = x + y

In[12] := **Simplify[(x + y)(x^2 - x y + y^2)]**

Out[12] = $x^3 + y^3$

Notice the space between **x** and **y** on the last input line. This space is necessary, because *Mathematica* would have interpreted **xy** as a new variable rather than the product of **x** and **y**. Even better, we could put an "∗" between **x** and **y**.

3.3.1   Assignments

When working with symbolic expressions, you often need to assign a numerical value to one (or more) of the original variables in the expression. In *Mathematica*, a single equal sign is used to assign values to a variable. For instance,

In[13] := **x = 5**

Out[13] = 5

will give the variable **x** the value 5 from now on. Whenever *Mathematica* sees **x**, it will substitute the value 5.

In[14] := **x^2 + 3 x∗y + y**

Out[14] = 25 + 16 y

To clear the variable **x**, type either **x = .** or **Clear[x]**.

Assignments can be quite general.

In[15] := **Clear[x]**

In[16] := **z = x^2 + 3 x∗y + y**

Out[16] = x^2 + y + 3 x y

In[17] := **z + 7 y**

Out[17] = x^2 + 8 y + 3 x y

In[18] := **y = 5**

Out[18] = 5

In[19] := **z**

Out[19] = 5 + 15 x + x^2

A variable or function name can be any string of letters and digits, as long as it begins with a letter, *never* a number. For example **x2** could be a variable, but **2x** will be interpreted as **2*x** by *Mathematica*. You should choose names that are both distinctive and easy to remember. For example, you might use **solndampedpend** as the name of the solution of a differential equation that models a damped pendulum.

A common source of puzzling errors in a *Mathematica* session is to forget that you have defined variables. *Mathematica* never forgets—when you assign a value to a particular variable, *Mathematica* assumes that you *always* want that variable to have that particular value, until you explicitly tell it otherwise (or start a new *Mathematica* session). You can check the current definition of a variable by typing the variable name by itself on an input line and evaluating it. For example, to find the current value of the variable **z**, type **z** and then SHIFT+ENTER. If *Mathematica* outputs **z**, it means that **z** has not been assigned a value. In some cases, you can get more information by typing **?** followed by the variable name.

3.3.2 Suppressing Output

Certain commands produce output that might be considered superfluous. For example, when you assign a value to a variable, *Mathematica* will echo the value. The output of a *Mathematica* command can be suppressed by putting a *semicolon* (**;**) after the command. Semicolons can also be used to separate a string of *Mathematica* commands when you are only interested in the output of the final command. Ending a statement with a semicolon has no effect on what *Mathematica* does internally; it only affects what you see in the Notebook.

3.3.3 Referring to Previous Output

Mathematica labels every input and every output; this makes it easy to use previous results in future calculations. The "**%**" symbol means "the most recent output", "**%%**" gives the next most recent result, **%%...%** (k times) gives the kth most recent result, and **%n** or **Out[n]** refers to the output labeled Out[n]. Here is an illustration:

```
In[20] := 2 + 2

Out[20] = 4

In[21] := %^2

Out[21] = 16

In[22] := %% * 5

Out[22] = 20

In[23] := %20/4

Out[23] = 1
```

A safer way to refer to past output is to assign the output to a variable. For example, we could type:

```
In[24]:= total = 2 + 2
```

```
Out[24]= 4
```

```
In[25]:= total^2
```

```
Out[25]= 16
```

The reason this is safer is that the output immediately above your current input in the Notebook is not necessarily the most recent output generated by *Mathematica*, for instance if you open more than one Notebook and switch between them. Furthermore, if you later re-evaluate a Notebook, the output numbering may be different.

3.4   Functions and Expressions

Mathematica allows you to use both built-in functions and functions you define yourself.

3.4.1   Built-in Functions

Mathematica has all of the usual "elementary functions" built in. For example, **Exp[x]** (*not* **e^x**, though you can type **e^x** using the special font *e* on the **BasicMathInput** palette) is the exponential function of **x**, and **Log[x]** (*not* **Ln[x]**) is the natural logarithm of **x**:

```
In[26]:= Log[10]//N
```

```
Out[26]= 2.30259
```

Note the use of square brackets instead of parentheses. Other built-in functions include **Sqrt**, **Cos**, **Sin**, **Tan**, and **ArcTan**. *Mathematica* also has a few built-in constants, such as **Pi** (the number π), **E** (the base *e* of the natural logarithm), **I** (the complex number $i = \sqrt{-1}$), and **Infinity** (∞). Essentially all functions and symbols with built-in meanings in *Mathematica* have names that start with capital letters.

3.4.2   User-defined Functions

It is possible to define new functions in *Mathematica*. In the following example, a polynomial function is defined and evaluated at one point.

```
In[27]:= f[x_] := x^3 + 7 x - 5
```

```
In[28]:= f[2]
```

```
Out[28]= 17
```

Notice the underscore (_) after the initial **x**, which means that **x** is a "dummy variable" in the definition, and the colon before the equals sign (**:=**), which tells *Mathematica* not to evaluate the right-hand side until you actually use the function.

3.4.3 Expressions

Another kind of object in *Mathematica*, similar to a function, is an *expression*. For example, **x^3 + 7 x - 5** is an expression. We can assign an expression to a variable, as follows:

```
In[29] := g := x^3 + 7 x - 5
```

Although this *expression* looks very much like the *function* we defined above, it is in fact quite different. The function is a rule for changing x into some other quantity, while the expression is merely a quantity involving x. In particular, typing **g[2]** will not give the value of the expression at $x = 2$. (Try it to see what happens!) We can, however, substitute a particular value into an expression by using the *replacement operator* **/.** to apply a transformation rule to the expression. A *transformation rule* is defined by a right arrow, typed as the combination **->**.

```
In[30] := g /. x → 2
```

```
Out[30] = 17
```

We will discuss the replacement operator and transformation rules further in Section 8.5.

3.5 Lists and Tables

Mathematica arranges its data in the form of lists. A list can contain any collection of *Mathematica* objects, including numbers, variables, functions, and equations. Lists are entered in the form **{a, b, c}**, where **a**, **b**, and **c** are the elements of the list. In many cases, lists can be treated exactly like single objects. For example, we could type

```
In[31] := v = {1, 2, 3, 4, 5}
```

```
Out[31] = {1, 2, 3, 4, 5}
```

```
In[32] := v^2
```

```
Out[32] = {1, 4, 9, 16, 25}
```

In effect, we instructed *Mathematica* to square the list **v**, and it responded by squaring each element of the list. You can also add, multiply, subtract, and divide lists, plot lists of functions, and solve lists of equations. You can even apply functions to lists, as in **f[v]**.

3.5.1 Extracting Parts of Lists

To extract the nth element of the list **v**, type **v[[n]]** or **Part[v, n]**. The following example gives the third element of **v**.

```
In[33] := v[[3]]
```

```
Out[33] = 3
```

A sublist of elements can be extracted from a list by specifying a list of positions. For example, to obtain a list consisting of the second, fourth, and fifth elements of **v^2**, type `(v^2) [[{2, 4, 5}]]` or

```
In[34]:= Part[v^2,{2, 4, 5}]
```

```
Out[34]= {4, 16, 25}
```

The **First** and **Last** commands can be used to extract the first and last elements of a list, respectively, as in **First[v]**. Here is the last element of **v^2**.

```
In[35]:= Last[v^2]
```

```
Out[35]= 25
```

As we shall see later, **First** is especially useful for extracting solutions to differential equations generated by the *Mathematica* commands **DSolve** and **NDSolve**.

3.5.2 Combining Lists

The **Join** command is used for combining lists. For example,

```
In[36]:= list1 = {a, b, c};
```

```
In[37]:= list2 = {c, a, d};
```

```
In[38]:= list3 = {a, d};
```

```
In[39]:= Join[list1, list2, list3]
```

```
Out[39]= {a, b, c, c, a, d, a, d}
```

To create a sorted list of the distinct elements of **list1**, **list2**, and **list3**, use the **Union** command.

```
In[40]:= Union[list1, list2, list3]
```

```
Out[40]= {a, b, c, d}
```

The **Riffle** command combines lists by interleaving the elements of the lists.

```
In[41]:= Riffle[{1, 2, 3}, {a, b, c}]
```

```
Out[41]= {1, a, 2, b, 3, c}
```

3.5.3 Tables of Values

An easy way to generate lists is to use *Mathematica*'s **Table** command. For example, to generate a table of values of the function **f** (defined above as $f(x) = x^3 + 7x - 5$) on the even integers from 0 to 10, you would type

```
In[42]:= Table[f[x], {x, 0, 10, 2}]
```

```
Out[42]= {-5, 17, 87, 253, 563, 1065}
```

In this example, the "**0**" represents the starting value of **x**, "**10**" is the ending value, and "**2**" is the increment (so we get only even integers in this case). The increment is an optional argument, the default increment being 1. The output of **Table** can be put in tabular form by appending the command **//TableForm**. In the following example, we make a table of values of **x** together with values of **f[x]**, for integer values of **x** from -1 to 3.

```
In[43]:= Table[{x, f[x]}, {x, -1, 3}]//TableForm
```

```
Out[43]//TableForm=
```

```
-1   -13
 0    -5
 1     3
 2    17
 3    43
```

In certain situations you may want to construct lists that depend on two or more parameters. You might do this, for example, if you're solving a second order differential equation in y, and want to study the solutions for various values of the initial conditions $y(x_0) = y_0$, $y'(x_0) = y'_0$. You can do this by using multiple parameters in the **Table** command. Here is an example in which we construct the set of all fractions with numerator and denominator between 1 and 3.

```
In[44]:= Table[a/b, {a, 1, 3}, {b, 1, 3}]
```

$$\text{Out[44]} = \left\{\left\{1, \frac{1}{2}, \frac{1}{3}\right\}, \left\{2, 1, \frac{2}{3}\right\}, \left\{3, \frac{3}{2}, 1\right\}\right\}$$

The table in this example is a *list of lists* or *nested list*. Note that there are two levels of braces, one for each parameter in the **Table** command. To remove the extra braces we can use the command:

```
In[45]:= Flatten[%]
```

$$\text{Out[45]} = \left\{1, \frac{1}{2}, \frac{1}{3}, 2, 1, \frac{2}{3}, 3, \frac{3}{2}, 1\right\}$$

Finally, to sort and eliminate all repetitions from this list we can type the **Union** command.

```
In[46]:= Union[%]
```

$$\text{Out[46]} = \left\{\frac{1}{3}, \frac{1}{2}, \frac{2}{3}, 1, \frac{3}{2}, 2, 3\right\}$$

3.6   Solving Equations

Before you solve differential equations with *Mathematica*, it is helpful to learn how to solve algebraic equations. For example, to solve the quadratic equation $x^2 + 2x - 4 = 0$ for the variable x, type:

```
In[47]:= Solve[x^2 + 2 x - 4 == 0, x]
```

$$\text{Out[47]} = \left\{\left\{x \to -1 - \sqrt{5}\right\}, \left\{x \to -1 + \sqrt{5}\right\}\right\}$$

Here the equation to be solved is specified using a double equal sign (==). The solutions are given as replacements for x. If the equation you are solving only has one variable, you can omit the final argument in the **Solve** command. For example, the syntax for solving $x^2 - 3x = -7$ is:

In[48]:= **Solve[x^2 - 3 x == -7]**

Out[48]= $\left\{\left\{x \rightarrow \frac{1}{2}\left(3 - i\sqrt{19}\right)\right\}, \left\{x \rightarrow \frac{1}{2}\left(3 + i\sqrt{19}\right)\right\}\right\}$

The answer consists of the exact complex number solutions $(3 \pm \sqrt{19}\,i)/2$, where the letter i (which *Mathematica* displays as **i**) in the answer stands for the imaginary unit $\sqrt{-1}$. To get numerical solutions, type **// N** after the input line.

The **Solve** command can solve higher-degree polynomial equations, as well as many other types of equations. It can also solve equations involving more than one variable. If there are fewer equations than variables, you should specify which variable(s) to solve for. For example, type **Solve[x + Log[y] == 3, y]** to solve $x + \ln y = 3$ for y in terms of x. *Mathematica* can also solve sets of simultaneous equations. You simply give the *list* of equations, and specify the list of variables to solve for. For example:

In[49]:= **Solve[{x + y^2 == 2, y - 3 x == 7}, {x, y}]**

Out[49]= $\left\{\left\{x \rightarrow \frac{1}{18}\left(-43 - \sqrt{157}\right), y \rightarrow \frac{1}{6}\left(-1 - \sqrt{157}\right)\right\},\right.$
$\left.\left\{x \rightarrow \frac{1}{18}\left(-43 + \sqrt{157}\right), y \rightarrow \frac{1}{6}\left(-1 + \sqrt{157}\right)\right\}\right\}$

This system of equations has two solutions. *Mathematica* reports its results as a list of transformation rules for x and y.

Some equations cannot be solved symbolically, and in these cases **Solve** will give you a warning message. For example, here is what happens when you try to evaluate **Solve[Cos[x] == x]**:

In[50]:= **Solve[Cos[x] == x]**

Solve::tdep : The equations appear to involve the variables
to be solved for in an essentially non-algebraic way. ≫

Out[50]= Solve[Cos[x] == x]

The graphs of $\cos(x)$ and x are shown in Figure 3.1; the intersection of the two curves represents the solution of the equation $\cos(x) = x$.

In this example, you can numerically find the (approximate) solution shown on the graph with the **FindRoot** command, which numerically searches for a solution of a given *equation*. So to find an approximate solution of the equation $\cos x = x$ near $x = 1$, type:

In[51]:= **FindRoot[Cos[x] == x, {x, 1}]**

Out[51]= $\{x \rightarrow 0.739085\}$

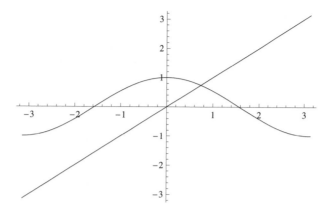

Figure 3.1: Two Intersecting Curves

3.7 Graphics

In this section, we introduce *Mathematica*'s basic plotting commands and show how to use them to produce a variety of plots.

3.7.1 Basic Plotting

The simplest way to graph a function of one variable is with *Mathematica*'s **Plot** command. The command **Plot[f[x], {x, x_{min}, x_{max}}]** will plot $f(x)$ as a function of x from x_{min} to x_{max}. If the function you want to plot is not already defined, you can type it in as an expression. For example, to graph x^2 on the interval -1 to 2, type:

```
In[52] := Plot[x^2, {x, -1, 2}]
```

Out[52] =

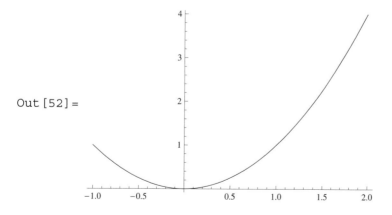

3.7.2   Modifying Graphs

The **Plot** command has many options that you can modify. Each time you produce a plot, you can specify the options you wish to modify by including a rule or sequence of rules in the form *optionname* → *value* as the last argument(s) to the **Plot** function. For example, to draw a frame around the preceding plot and add a title, type:

```
In[53]:= Plot[x^2, {x, -1, 2}, Frame → True,
            PlotLabel → "The Parabola" y == x^2]
```

Out[53]=

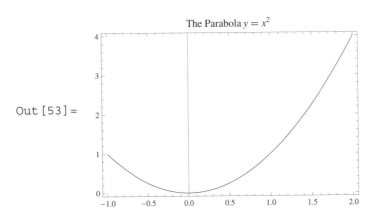

You can add labels on the horizontal and vertical axes with the **AxesLabel** option. The expressions you give as labels are printed just as they would be if they appeared as *Mathematica* output. You specify arbitrary text by putting it inside a pair of double quotes.

When *Mathematica* produces a plot, it automatically sets the horizontal and vertical scales to include what it interprets as the "interesting" parts of the plot. You can change the horizontal and vertical ranges of the graph with the **PlotRange** option. For example, to confine the vertical range to the interval from 0 to 3 in the example above, add the option **PlotRange → {0, 3}**. For other possibilities, see the online help for **PlotRange**. When no explicit limits are given for a particular coordinate, *Mathematica*'s default settings are automatically applied.

By default, *Mathematica* scales the axes so that the ratio of width to height in the plot is equal to the inverse of the "golden ratio" (approximately 1.618)—supposedly the most pleasing shape for a rectangle. You can adjust the scaling with the **AspectRatio** option; for example, you can make the plot square by setting **AspectRatio → 1**.

3.7.3   Plotting Multiple Curves

Mathematica's **Plot** command can also plot several functions together on the same set of axes. You simply specify the functions as a list in the **Plot** command. *Mathematica* will choose a different color for each function automatically. For example, the command

`Plot[{Cos[x], x}, {x, -Pi, Pi}]` will produce the graphs shown in Figure 3.1 above.

Another way to plot several functions is to plot them separately and assign names to the plots. The plots then can be combined and displayed using the **Show** command. All of the options for the resulting graphic will be based on the options of the first graphic listed in the **Show** expression. For example, the following commands will also produce Figure 3.1.

```
In[54]:= plot1 = Plot[Cos[x], {x, -Pi, Pi}];
```

```
In[55]:= plot2 = Plot[x, {x, -Pi, Pi}];
```

```
In[56]:= Show[plot1, plot2, PlotRange → {-Pi, Pi}]
```

Notice that we used the option **PlotRange → {-Pi, Pi}** to expand the vertical range on the resulting graphic to $[-\pi, \pi]$. This was necessary because the vertical range for **plot1** was set to $[-1, 1]$ by *Mathematica*, and **plot1** was displayed first. Alternatively we could have typed **Show[plot2, plot1]** to produce the same graphic, since the vertical range for **plot2** is $[-\pi, \pi]$.

3.7.4   Plotting Data Lists

Mathematica can create plots from lists of data as well as functions. The *Mathematica* command for plotting lists is **ListPlot**. **ListPlot**'s commands are analogous to those for plotting functions. **ListPlot[{{x₁, y₁}, {x₂, y₂}, ...}]** plots a list of points with specified x and y coordinates. **ListPlot[{y₁, y₂, ...}]** will plot points at $y_1, y_2, \ldots$ corresponding to x coordinates assumed to be $1, 2, \ldots$. The list may be given explicitly or it may be generated by the **Table** command. For example,

```
In[57]:= table1 = Table[i^2, {i, 10}];
```

```
In[58]:= ListPlot[table1]
```

Out[58]=

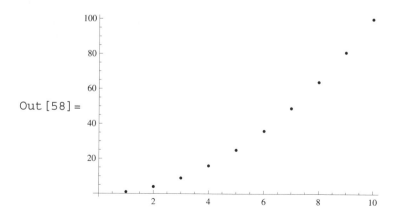

To connect the plotted points with line segments, use the command **ListLinePlot** in place of **ListPlot**. Alternatively, you could use the option **Joined → True** in the **ListPlot** command.

3.7.5 Parametric Plots

Mathematica can also produce parametric plots. The command **ParametricPlot [{x[t], y[t]}, {t, t$_{min}$, t$_{max}$}]** plots the parametric curve defined by $x(t)$ and $y(t)$ as the parameter t varies over the interval from t_{min} to t_{max}. Here is an example:

```
In[59]:= x[t_] := Exp[-t/100]*Cos[t]
```

```
In[60]:= y[t_] := Exp[-t/100]*Sin[t]
```

```
In[61]:= ParametricPlot[{x[t],y[t]}, {t, 0, 100}]
```

Out[61]=
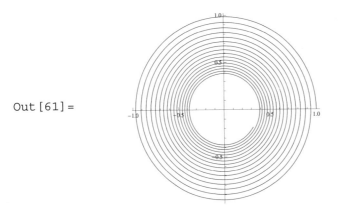

Note that we first defined x and y as functions of t and then we applied **ParametricPlot** to the expressions **x[t]** and **y[t]**. Alternatively, we could have entered the expressions for $x(t)$ and $y(t)$ directly into the **ParametricPlot** command. If you want to apply **ParametricPlot** (or any plotting commands) to a previously defined function **f**, you must explicitly enter the independent variables, as in **f[x]**, rather than just **f**. Another way of saying this is that the plotting routines plot *expressions*, and not *functions*.

Parametric plots will be used in Chapters 8 and 12 to produce phase portraits of systems of differential equations.

3.7.6 Contour Plots

A contour plot of an expression in two variables is a plot of the *level curves* of the expression, *i.e.*, sets of points in the x-y plane where the expression has constant value. For example, the level curves of the expression $x^2 + y^2$ are circles, and the *levels* are the squares of the radii of the circles. Contour plots are produced in *Mathematica* with the **ContourPlot** command. To make a contour plot of $x^2 + y^2$ type:

```
In[62]:= ContourPlot[x^2 + y^2, {x, -3, 3}, {y, -3, 3}]
```

Out[62]=

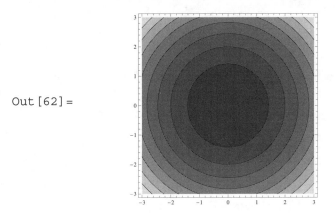

The shading in the plot represents different values of the function: the larger the value of the function, the lighter the color. (The default color scheme uses shades of blue and purple; since this book is printed in black and white, we have changed it to use shades of gray with the option **ColorFunction → "GrayTones"**.) Shading can be turned off using the option **ContourShading → False**. You can specify *how many* level sets to plot by adding the **Contours** option: **Contours → n** produces n contours. The option **Contours → {a, b, ...}** produces contours at levels $a, b, \ldots$. For example, to plot the circles of radius 1, $\sqrt{2}$ and $\sqrt{3}$, type:

```
In[63]:= ContourPlot[x^2 + y^2, {x, -3, 3}, {y, -3, 3},
           Contours → {1, 2, 3}, ContourShading → False]
```

Contour plots will be used in Chapter 5 to plot implicit solutions of differential equations.

3.8 Calculus

The **D** command differentiates a symbolic expression, and the **Integrate** command performs integration. For example, to differentiate x^n, you could type:

```
In[64]:= D[x^n, x]
```

Out[64]= $n\, x^{-1+n}$

To integrate $\cos x$ type:

```
In[65]:= Integrate[Cos[x], x]
```

`Out[65]= Sin[x]`

This is an example of an indefinite integral. You can also compute a definite integral as follows:

`In[66]:= Integrate[x^2*Exp[-x], {x, -5, 5}]`

$$Out[66]= -\frac{37}{e^5} + 17\,e^5$$

Some integrals cannot be done symbolically, and in these cases *Mathematica* returns the unevaluated integral:

`In[67]:= hard = Integrate[Log[1 + x^2]*Exp[-x^2], {x, 0, 1}]`

$$Out[67]= \int_0^1 e^{-x^2}\,\mathrm{Log}\!\left[1 + x^2\right]\,dx$$

In that case you can use *Mathematica* to compute a numerical approximation to the definite integral in one of two ways. First, you can use **N**:

`In[68]:= N[hard]`

`Out[68]= 0.153886`

Second, *Mathematica* has a built-in numerical integration routine: **NIntegrate[f[x], {x, x$_{\text{min}}$, x$_{\text{max}}$}]** gives a numerical approximation to the integral $\int_{x_{min}}^{x_{max}} f(x)\,dx$.

`In[69]:= NIntegrate[Log[1 + x^2]*Exp[-x^2], {x, 0, 1}]`

`Out[69]= 0.153886`

Mathematica can also deal with improper integrals. For instance, you can integrate to ∞, which is written **Infinity** in *Mathematica*:

`In[70]:= Integrate[Sin[x]/x, {x, 0, Infinity}]`

$$Out[70]= \frac{\pi}{2}$$

Even integrals with singularities can often be evaluated:

`In[71]:= improper = Integrate[Log[ArcTan[x]], {x, 0, 1}]`

$$Out[71]= \int_0^1 \mathrm{Log}[\mathrm{ArcTan}[x]]\,dx$$

`In[72]:= N[improper]`

`Out[72]= -1.09053`

The command **NIntegrate[Log[ArcTan[x]], {x, 0, 1}]** yields the same answer.

3.9 Packages

One of the most important features of *Mathematica* is that it is an extensible system. There is a great deal of mathematical and other functionality that is built into *Mathematica*; however, there are many specialized functions that are not loaded when *Mathematica* is initially started. By loading *Mathematica* packages, it is possible to extend *Mathematica*'s functionality. To use commands in packages, you must explicitly tell *Mathematica* to load the commands by typing `<< package`. For example, the `VectorFieldPlot` command is part of the `VectorFieldPlots` package, and must be loaded before it can be used. To load this package, type

```
In[73]:= << VectorFieldPlots`
```

Note the character at the end of the name of the package is a single *backquote*, or *grave accent*. You will need the `VectorFieldPlots` package to do some of the problem sets. It contains a collection of specialized routines for graphing vector fields.

If you use a package command without loading the package first, *Mathematica* will not recognize the command and will echo the input line in the output area. If you do make this mistake, you can recover by typing `Remove[command]` and then loading the package properly. For example, if we had tried to use `VectorFieldPlot` before loading the `VectorFieldPlots` package, we could type `Remove[VectorFieldPlot]` before loading the package and reusing the command.

3.10 Some Tips and Reminders

We have introduced quite a few commands and concepts in this chapter. It will take you a while to become comfortable and proficient with them. In particular, the distinctions between functions and expressions, and between `=` and `:=`, are often quite subtle and can trip up even experienced users. The best way to learn is through experience and by using on-line help (see Section 2.5) to access *Mathematica*'s Documentation Center when necessary.

Keep in mind that previously defined (and forgotten) variables are often the source of mysterious errors. To avoid mistakes, you should remove values you have defined as soon as you have finished using them. Use `Clear` to clear old variables and functions. Also, remember that built-in *Mathematica* objects always have names starting with upper-case letters. To avoid confusion, you should always choose names for your own variables that start with lower-case letters. Finally, remember that typing ALT+. will usually abort a *Mathematica* command. If *Mathematica* gets hung up, try quitting the kernel to end the session.

Chapter 4

Using *Mathematica* Notebooks

In this chapter, we describe some aspects of the Notebook feature of *Mathematica*. Our comments apply to *Mathematica* 6 (many aspects are different in earlier versions) running on a Windows platform. However, most of the Notebook features are the same on all platforms. In particular, we describe some of the formatting tools that will help you create an attractive and polished document when you prepare solutions to the problem sets. You can refer to the *Sample Solutions* at the end of this book to see additional examples of formatted *Mathematica* documents.

Since we refer frequently to menu items, we establish the following convention. When referring to an item in a submenu of another menu, we type the name of the top-level menu, then a black triangle, then the name of the item or submenu, and so on. For example, **Format ▶ Face ▶ Italic** means the **Italic** item of the **Face** submenu of the **Format** menu. This item is used to change text into or out of italics.

Remark 4.1 You might have noticed that some menu items have not only a name but also some extra symbols on the right. These symbols are the *keyboard shortcut* for the given menu item. You might find these keyboard shortcuts to be efficient alternatives to using the mouse. For example, the **Format ▶ Face ▶ Italic** item is followed by "CTRL+I".

4.1   The Kernel and the Front End

The *Mathematica* system consists primarily of two parts: the kernel and the front end. The kernel is the computational engine of *Mathematica*. It is the part that actually does the calculations, generates the In and Out labels, and handles various computational tasks for formatting the output.

The front end is the part of *Mathematica* that interacts with the kernel. It reads input, displays output, and manipulates Notebooks. The Notebook is the medium you will use to interact with *Mathematica*.

To evaluate your instructions, type the instructions and then press SHIFT+ENTER or the ENTER key on the numeric keypad. Keep in mind when using *Mathematica* that there is a difference between typing the ENTER key on an alphanumeric keyboard (just above the right SHIFT key) and typing SHIFT+ENTER (holding down the SHIFT key and pressing EN-TER). Typing SHIFT+ENTER tells *Mathematica* to evaluate the instruction you've typed. The ENTER key is a simple linefeed; it generates a new line without causing *Mathematica* to evaluate the input. It is useful if you want to group several commands or expressions together in a single *Input* cell; see Section 4.2.2 for more about cells. In this case, when you evaluate the Input cell, all of the commands will be executed in order, and their results will appear in separate Output cells. (On some systems, the ENTER key is labeled RETURN. On all systems, the ENTER key on the numeric keypad has the same effect as SHIFT+ENTER.)

After the front end sends your input to the kernel, but before the result is returned to the front end, the cell bracket on the right side of the Notebook window will be highlighted, and the word "Running..." will appear in the title bar of the Notebook. If you are evaluating your first input cell during a *new Mathematica* session, there will be a brief delay (typically a few seconds) as the front end automatically starts the kernel. From this point, responses to simple inputs should be nearly instantaneous. Note that you will not see any In or Out labels until results are returned from the kernel.

4.2   The Notebook

All of your interactions with *Mathematica*'s front end are entered and displayed in a *Mathematica* Notebook. A Notebook appears on your computer screen as a window with various kinds of text and graphics in it. *Mathematica* treats a Notebook as a *document*, and in particular, since the front end supports all the usual style mechanisms available in word processors, you can use the front end to produce a nicely formatted version of the Notebook. Typically you will do this by choosing appropriate items from the **Format** menu.

You can open more than one Notebook at a time by clicking on **File ▶ New ▶ Notebook (.nb)** (or typing CTRL+N) to start a new Notebook. To open a previously saved Notebook, select **Open** (or type CTRL+O). To close the active Notebook without quitting *Mathematica*, select **Close** (or type CTRL+F4) and to close all open Notebooks without quitting *Mathematica* select **Close All** (or type SHIFT+CTRL+F4). If you try to close an unsaved Notebook, *Mathematica* will ask whether you want to save it first.

The active Notebook is the one with the darkened titlebar, and anything you type will go into that Notebook. You can select a Notebook by clicking anywhere in its window. To move a Notebook, drag its titlebar.

One reason to open more than one notebook is to copy and paste text from one Notebook to another. Note, however, that any definitions of variables, *etc.*, that you make in one Notebook carry over to every other open Notebook, so if you evaluate Input cells in two different Notebooks, you may get unexpected results.

Save your Notebooks often. The simplest way to save is to click on **File ▶ Save** or type CTRL+S.

4.2.1 Quitting the Kernel

Sometimes you may want to start a fresh *Mathematica* session without exiting the program, restarting, and reopening the Notebook(s) you are working on. This might happen, for example, if some previously defined variables or functions are causing problems that you can't easily eradicate. Or you may simply want to reset the In and Out numbering. One way to take care of these problems is to quit and restart the *Mathematica* kernel. When you quit the kernel, the underlying *Mathematica* process will die, along with any variable or function definitions, but your Notebooks will remain on the screen. The next time you press SHIFT+ENTER, a new kernel will start up, and after a slight delay you can continue working as before. The In and Out numbering will be initialized to 1.

To quit the kernel, you can select **Evaluation ▶ Quit Kernel ▶ Local**. (In networked configurations, there may be other choices besides **Local**; you will then have to figure out which one is applicable in your situation.) A box will appear, asking you if you really want to quit the kernel; click on the appropriate button to confirm your choice.

4.2.2 Cells

A *Mathematica* Notebook is divided into *cells*, which are delineated by a bracket along the right-hand margin of the Notebook. Each cell within a Notebook is assigned a particular style that indicates its role within the Notebook. There are several different styles of cells, including *Input*, *Output*, *Text*, *Title*, and *Section* styles. Input cells, where you type expressions to be evaluated by the *Mathematica* kernel, are typically in Input style, Output cells, where *Mathematica* gives its responses, are typically in Output style, and cells that contain text that is intended purely to be read are typically in Text, Title, or Section style. For example, to make and interpret a graph, you would first type a plotting command in an Input cell, then *Mathematica* would create the graph in an Output cell, and finally you would describe it in a Text cell.

Each cell style defines various settings for the options associated with a cell, such as font, color, justification, alignment, and so on. The easiest way to change the style of a cell is to select the cell by clicking on its bracket (the bracket will be highlighted), and then select a style from the **Format ▶ Style** menu or type a keyboard shortcut to change the style of the cell. For example, to change a cell to a Text cell, click on the cell bracket, and then type ALT+7.

4.2.3 Cell Hierarchy

The hierarchy of cells serves as a structure for organizing the information in a Notebook, as well as specifying the overall look of the notebook. A Notebook is organized by grouping together blocks of cells. An example of this is the automatic grouping of Input cells with Output cells. A cell grouping is indicated by nested brackets along the right edge of the Notebook. When you use "sectioning" cells, the remaining cells are automatically grouped according to a *cell hierarchy*. For example, if you create a Title cell at the top of your Notebook, there will be an infinitely expandable cell bracket at the far right of the Notebook that will enclose every other cell you create in the Notebook. If you then create a Section cell, there will be a new expandable bracket created just inside the first one, which will surround all successive cells until you create another Section cell. The same is true for Subsection and Subsubsection cells.

Here is an example of how this can be useful. When you start the solution to a problem, say Problem 1, create a Section cell and type "Problem 1". Now create your solution to Problem 1 using Text cells, Input cells, Output cells, *etc.* All of these cells will be grouped together with a bracket along the right side of the Notebook. If Problem 1 has parts (a), (b), *etc.*, then use a Subsection cell to denote the beginning of each part. When you are finished with Problem 1, create a new Section cell and type "Problem 2". A new grouping bracket for Problem 2 will be created, and subsequent material in the Notebook will be grouped in this bracket until a new sectioning cell is introduced.

These grouping brackets can be used to manipulate entire blocks of cells in the Notebook. For example, if you have created a section that contains the solution to Problem 1, you can click on its grouping bracket and the bracket will be highlighted. If you press SHIFT+ENTER, then *Mathematica* will evaluate all the Input cells in the section, in order. If you double-click on its grouping bracket, the entire section will close up, and only the section title will be visible.

When a cell group is closed, only the first cell in the group is displayed by default. A small arrow at the bottom of the bracket indicates that its cell group is closed. To open the group, just double-click on its bracket once more. Double-clicking the bracket of a cell that is not the first of a cell group closes the cell group around that cell and creates a bracket with up and down arrows (or only an up arrow if the cell was the last in the group).

If you experiment with this feature, you'll soon see how useful it is for organizing and expediting your work. Note that when you print the Notebook, it will appear much like it does on the screen. In particular, closed cell groups will print in closed form.

4.2.4   Pointers and Insertion Points

When *Mathematica* is first started, it displays an empty Notebook, with the mouse pointer in the shape of a horizontal I-beam. You can start typing right away. The standard text insertion point for editing text will appear after you begin typing.

A cell insertion point is indicated by a solid black line running horizontally across the Notebook. This horizontal line is a *cell insertion bar* and appears at the boundary between two cells or at the beginning or end of the Notebook. To insert a new cell, move the mouse pointer in the Notebook window until it becomes a horizontal I-beam. Click the mouse button, and a cell insertion bar will be inserted into the Notebook. Choose an item from the **Format ▶ Style** menu or type a keyboard shortcut to select a cell style, and then start typing. If you do not choose a cell style prior to typing, new cells will be Input cells by default. For example, to insert a new Text cell at the end of a Notebook, you click at the end of the Notebook, type ALT+7 or select **Format ▶ Style ▶ Text**, and then start typing.

To edit text in a cell, simply click in the cell to position the standard text insertion point, where you can add text by typing, or delete text using the backspace or delete key. You can also use the arrow keys on your keyboard to move the insertion point around. Note that the insertion point, *i.e.*, the place where text appears, is generally different from the mouse pointer position.

4.2.5 Manipulating Cells

Occasionally, you will want to divide or merge cells. To divide a cell, first put the insertion point where you want the division to occur, and then select **Cell ▶ Divide Cell** or type SHIFT+CTRL+D. To merge several contiguous cells, first select the cells by dragging the mouse over a range of cell brackets, and then select **Cell ▶ Merge Cell** or type SHIFT+CTRL+M. The selected cells will merge, and the cell style of the new cell will be the style of the topmost cell in the selection. If the selection contains a mixture of text and mathematical formulas or graphics cells, the result will be a text cell with in-line mathematical formulas and/or graphics cells. Formatted cells generally should not be merged or divided.

The menu command **Insert ▶ Input from Above** (CTRL+L) is useful for manipulating input. This menu command places a copy of the contents of the nearest Input cell preceding your insertion point at the current insertion point. This is especially useful if you want to execute several variants of a single input expression.

4.2.6 Mathematical Typesetting

When inserting commentary in Notebooks it is often useful to be able to type mathematical formulas, symbols, subscripts, *etc.*, and to use different fonts in some cases. You can do these things through keyboard sequences, menu items, and *palettes*. The "Basic Math Input" palette is the thin window that generally appears automatically when you start *Mathematica*; if no such window appears for you, then select **Palettes ▶ BasicMathInput** to open it. Other palettes are also available in the **Palettes** menu, but we will concentrate on the Basic Math Input Palette, which offers common mathematical symbols, Greek letters, and templates for constructing formulas. Click on a button to insert the corresponding symbol, letter, or template into the active Notebook.

To type a formula in a Text cell, start by typing CTRL+9 (this is the keyboard shortcut for **Insert ▶ Typesetting ▶ Start Inline Cell**). Then type the formula using the keyboard and the palette. Many forms of mathematical notation, such as subscripts, superscripts, fractions, and square roots, can be entered either from the keyboard or through a template from the palette. For example, to type x^2 you can type **x**, then CTRL+6, then **2**, then CTRL+SPACE to move the insertion point out of the superscript position down to the baseline. Alternatively, click on the superscript template in the upper left corner of the Basic Math Input palette. A pair of boxes appear, waiting for you to type the appropriate symbols into them. Type **x**, then press TAB (or click in the superscript box with the mouse), then type **2**, and finally type CTRL+SPACE. Then continue typing your formula. When you are done with the formula, type CTRL+0 (for **Insert ▶ Typesetting ▶ End Inline Cell**), and continue typing text.

You can learn the keyboard shortcuts for many common mathematical notations from the **Insert ▶ Typesetting** menu. For more information on mathematical typesetting, click on Math Typesetting in the Notebooks and Documents section of the Documentation Center (**Help ▶ Documentation Center**).

To change the font in a Text cell, use the items in the middle of the **Format** menu. For instance, to change a word, phrase, or entire cell into italics, select the text that you want to change (or select the cell). Then select **Format ▶ Face ▶ Italic** or type its keyboard shortcut, CTRL+I. You can also use this command to switch into and out of italics as you are typing a Text cell. Similarly, you can change the size of the font, its color, *etc.*

4.2.7   Displaying and Printing *Mathematica* Notebooks

There are several ways to customize the overall appearance of your *Mathematica* Notebooks. First we discuss ways to change how Notebooks are printed. One way is by using the items in the **File ▶ Printing Settings** menu. For example, you can adjust the margins by selecting **File ▶ Printing Settings ▶ Printing Options...**. Select **File ▶ Printing Settings ▶ Headers and Footers...** to change what is printed at the top and bottom of each page. And in Windows, you can set options specific to your printer by selecting **File ▶ Printing Settings ▶ Page Setup...**, and by clicking on "Properties" in the **Print...** dialog box.

To change how your Notebook looks overall, you can use a *stylesheet*. Stylesheets define a set of cell styles to be used in the Notebook. By changing a cell's style instead of assigning specific options to a cell, you apply all the cell attributes desired in one step and give your notebooks a consistent design. You can either select a predefined stylesheet from *Mathematica*'s collection of *stylesheets*, which are listed in the **Format ▶ Stylesheet** menu, or you can create your own stylesheet using **Format ▶ Edit Stylesheet...**. In general, the "Default" stylesheet used by *Mathematica* will be sufficient for your needs. To learn more about stylesheets, see *Mathematica*'s Documentation Center (**Help ▶ Documentation Center**).

The most efficient way to change the appearance of your Notebook on the screen or the way it is printed is to change the *style environment* for a particular stylesheet. Style Environments allow you to use one stylesheet in several different ways. For example, the 'Default' stylesheet used by *Mathematica* has the following style environments:

- Working — designed for on-screen use;

- Presentation — designed for presentations, either from a computer or printed on transparencies;

- SlideShow — designed for presentation slides;

- Condensed — designed for viewing on a small screen or printing on fewer pages;

- Printout — designed for optimal printing of the Notebook.

You can set the style environment used for printing or for on-screen use. To set the screen environment, go to **Format ▶ Screen Environment** and choose from the available environments listed. To set the printing environment, go to **File ▶ Printing Settings ▶ Printing Environment**. This allows you to display a notebook in one environment but automatically get printouts in another environment. For example, the 'Default' stylesheet is set up to use the "Working" environment onscreen and the "Printout" environment when printing. Because of this, the print version of a notebook will look slightly different from the onscreen version. If you want the print and screen versions to be identical, the screen and printing environments must be the same.

To uniformly change the font in a cell or group of cells, use the items in the middle of the **Format** menu. For instance, to increase the font size for an entire cell or group of cells, select the cell or group of cells that you want to change. Then select **Format ▶ Size ▶ Larger** or type its keyboard shortcut ALT+=. Similarly, you can change the face or color of the font or change the background color of a cell or group of cells.

Finally, you can change a large number of advanced formatting and printing options through the *Option Inspector*. To open the Option Inspector, select **Format ▶ Option Inspector....** For instructions on how to use it, see *Mathematica*'s Documentation Center.

4.2.8 Preparing Homework Solutions

You will use *Mathematica* to prepare solutions to the problem sets in this book. You should make use of the editing features of the *Mathematica* Notebook interface to produce a polished document. In particular, material that is not relevant to the final answer (typing errors, trial calculations, *etc.*) should be deleted from the final printed form of the Notebook. Answers to interpretive questions should appear in a logical place relative to the *Mathematica* output.

As an illustration, we present a *Mathematica* Notebook solution to Problem 2 of Problem Set A. We've used the "Default" stylesheet with the "Working" printing environment. Note also that since our output is intended for a black-and-white printer, we've used the **Plot** option **PlotStyle→Black** to override *Mathematica*'s coloring scheme for graphs.

Solution to Problem Set A, Problem 2

■ Contents

- Numerical Solution

- Graphical Solution

■ Numerical Solution

Evaluate to 15 digits: $\dfrac{\sin(x)}{x}$ for $x = 0.1, 0.01, 0.001$.

By default, *Mathematica* will use machine precision for all numerical approximations and then display its results to 6 significant figures. To avoid rounding off to machine precision, we specify the x values as a list of rational numbers:

```
In[1]:=   x = {1/10, 1/100, 1/1000};
```

Next we compute the desired values and use **N** to display them to 15 significant figures.

```
In[2]:=   y = N[Sin[x]/x, 15]

Out[2]=   {0.998334166468282, 0.999983333416666,
           0.999999833333342}
```

To view the values in table form, first use the **Riffle** command to interleave the elements of x and y, and then use the **Partition** command to partition the new list into blocks of 2.

```
In[3]:=   list1 = Riffle[x, y];
          list2 = Partition[list1, 2];
          TableForm[list2, TableHeadings → {None, {x, y}}]

Out[5]//TableForm=
```

x	y
0.1	0.998334166468282
0.01	0.999983333416666
0.001	0.999999833333342

These values illustrate the fact that the limit of $\sin(x)/x$ as x approaches 0 is 1.

■ Graphical Solution

We can also illustrate the same fact graphically.

```
In[6]:=   plot1 = ListPlot[list2,
             PlotStyle → {Black, PointSize[0.02]}];
          plot2 = Plot[Sin[x]/x, {x, 0, 0.1}
             PlotStyle → Black];
```

```
Show[plot1, plot2, PlotRange → {0.9982, 1},
    Axes → False, Frame → True,
    PlotLabel → "The Limit of sin(x)/x"]
```

Out [8] =

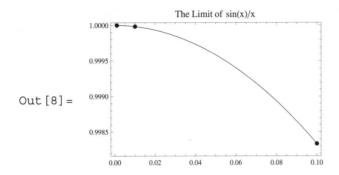

Problem Set A

Practice with *Mathematica*

In this problem set, you will use *Mathematica* to do some basic calculations, and then to plot, differentiate, and integrate various functions. This problem set is the minimum you should do in order to reach a level of proficiency that will enable you to use *Mathematica* throughout the course. A solution to Problem 2 appears in Chapter 4, and a solution to Problem 4 appears in the *Sample Solutions*.

1. Evaluate:

    (a) $\dfrac{413}{768 + 295}$ (as a decimal),

    (b) 2^{123}, both as an approximate value in scientific notation and as an exact integer,

    (c) π^2 and e to 35 digits,

    (d) the fractions $\frac{61}{88}$, $\frac{13863}{20000}$, and $\frac{253}{365}$, and determine which is the best approximation to $\ln(2)$.

2. Evaluate to 15 digits:

    (a) $10 \sin(1/10)$,

    (b) $100 \sin(1/100)$,

    (c) $1000 \sin(1/1000)$.

3. Graph the equations:

    (a) $y = x^3 - x$ on the interval $-1.5 \le x \le 1.5$,

    (b) $y = \tan x$ on the interval $-2\pi \le x \le 2\pi$,

    (c) $y^2 = x^3 - x$ on the interval $-2.5 \le x \le 2.5$. (*Hint*: Use `ContourPlot`.)

4. *Mathematica* can factor polynomials as well as integers. Look up the syntax and solve these problems.

    (a) Factor $x^3 + 5x^2 - 17x - 21$.

    (b) Find the prime factorization of 123456789.

5. Plot the functions x^8 and 4^x on the same graph, and determine how many times their graphs intersect. (*Hint*: You will probably have to make several plots, using various intervals, in order to find all the intersection points.) Now find the values of the points of intersection using the **FindRoot** command.

6. Compute the following limits:

    (a) $\displaystyle\lim_{x\to 0} \frac{\tan x}{x}$,

    (b) $\displaystyle\lim_{x\to 0^+} \frac{1}{x}$ and $\displaystyle\lim_{x\to 0^-} \frac{1}{x}$,

    (c) $\displaystyle\lim_{x\to\infty} x e^{-x^2}$ and $\displaystyle\lim_{x\to-\infty} x e^{-x}$,

    (d) $\displaystyle\lim_{x\to 0} \frac{\ln(1-x)+x}{x^2}$.

7. Compute the following derivatives:

    (a) $\displaystyle\frac{d}{dx}\left(\frac{x^3}{x^2+1}\right)$,

    (b) $\displaystyle\frac{d}{dx}\left(\sin\left(\sin\left(\sin x\right)\right)\right)$,

    (c) $\displaystyle\frac{d^3}{dx^3}\left(\arctan x\right)$,

    (d) $\displaystyle\frac{d}{dx}\left(\sqrt{1+x^2}\right)$,

    (e) $\displaystyle\frac{d}{dx}\left(e^{x\ln(x)}\right)$.

8. Compute the following integrals:

    (a) $\int e^{-3x}\sin x\, dx$,

    (b) $\int (x+1)\ln x\, dx$,

    (c) $\displaystyle\int_0^1 \sqrt{\frac{x}{1-x}}\, dx$,

    (d) $\displaystyle\int_{-\infty}^{\infty} e^{-x^2}\, dx$,

    (e) $\displaystyle\int_0^1 \sqrt{1+x^4}\, dx$. For this example, also compute the numerical value of the integral.

9. Use **Solve** to solve the equation

$$x^5 - 3x^2 + x + 1 = 0. \tag{A.1}$$

Find the numerical values of the five roots. Plot the graph of the 5th degree polynomial on the left-hand side of (A.1) on the interval $-2 \leq x \leq 2$. Now explain your results—in particular, reconcile your five roots with the fact that the graph touches the x-axis only twice. To verify this, you should restrict the y-axis appropriately.

10. In one-variable calculus you learned that the local maxima and minima of a differentiable function $y = f(t)$ are found among the *critical points* of f, that is, the points t where $f'(t) = 0$. For example, consider the polynomial function

$$y = f(t) = t^6 - 4t^4 - 2t^3 + 3t^2 + 2t$$

on the interval $[-3/2, 5/2]$.

(a) Graph $f(t)$ on that interval.

(b) How many local maxima and minima do you see? It's a little hard to determine what is happening for negative values of t. If need be, redraw your graph by restricting the t and/or y axis.

(c) Now use *Mathematica* to differentiate f and find the points t where $f'(t) = 0$ on the interval $[-3/2, 5/2]$. How many are there? Use **FindRoot** to hone in on their values.

(d) Now verify that the negative critical point is indeed an *inflection point* by graphing $f''(t)$ on the interval $-1.2 \leq t \leq -0.8$. How does that graph establish that the point is an inflection point?

11. (a) Use **Solve** to simultaneously solve the pair of equations

$$\begin{cases} x^2 - y^2 = 1 \\ 2x + y = 2. \end{cases}$$

(b) Plot the two curves on the same graph and visually corroborate your answer from part (a). (*Hint*: You can combine multiple plots using the command **Show**; see the online help for more information.)

12. Use **Sum** or the summation template in the **BasicMathInput** palette to do the following problems.

(a) Sum the famous infinite series $\sum_{n=1}^{\infty} \frac{1}{n^2}$.

(b) Sum the geometric series $\sum_{n=0}^{\infty} x^n$.

(c) Define a function $f(x) = \sum_{n=1}^{\infty} \frac{x^n}{n}$. Sum the series to get a simple expression for $f(x)$.

(d) Compute $f'(x)$ by differentiating the function you found in part (c).

(e) Now sum the series $\sum_{n=1}^{\infty} \frac{d}{dx}(\frac{x^n}{n})$. What does this suggest about differentiating a function that is defined by a power series?

13. (a) Use **ContourPlot** to plot the curve defined by the equation $3y+y^3-x^3 = 5$.

(b) Now plot the level curves of $3y + y^3 - x^3$ for the levels $-2, 0, 2, 5, 8$.

(c) Plot the level curve of the function $f(x, y) = y \ln x + x \ln y$ that passes through the point (1,1).

14. Consider the expression $u = e^{-x} \sin x + e^x \cos x$.

(a) We want the values of this expression at the points

$$-\pi, -\frac{3}{4}\pi, -\frac{1}{2}\pi, -\frac{1}{4}\pi, 0, \frac{1}{4}\pi, \frac{1}{2}\pi, \frac{3}{4}\pi, \pi.$$

First, use the command **Table** to create a list of these values and name it **xpts**.

(b) Use the replacement operator and a transformation rule to create a list of the values of u at the points **xpts**.

(c) Now define the function

$$f(x) = e^{-x} \sin x + e^x \cos x.$$

(d) Find a list of the values of f at the points **xpts**. Convert these values to decimal form with five digits.

(e) Use the command **TableForm** to display a table of the points **xpts** in the first column and the corresponding five-digit values of f in the second.

Chapter 5

Solutions of Differential Equations

In this chapter, we show how to solve differential equations with *Mathematica*. For many differential equations, the command **DSolve** produces the general solution to the differential equation, or the specific solution to an associated initial value problem. We also discuss the existence, uniqueness, and stability of solutions of differential equations. These are fundamental issues in the theory and application of differential equations. An understanding of them helps in interpreting and using results produced by *Mathematica*.

5.1 Finding Symbolic Solutions

Consider the differential equation

$$\frac{dy}{dt} = f(t, y). \tag{5.1}$$

A solution to this equation is a differentiable function $y(t)$ of the independent variable t that satisfies $y'(t) = f(t, y(t))$ for all t in some interval. For some functions f, it is possible to find a formula for the solutions to (5.1); we call such a formula a *symbolic solution* or *formula solution*. Finding a symbolic solution is generally not a straightforward task, and not surprisingly, computational algorithms for solving differential equations symbolically are imperfect. Nonetheless, *Mathematica*'s symbolic differential equation solver **DSolve** can correctly solve most of the differential equations that can be solved with the standard solution methods one learns in an introductory course.

A symbolic solution to (5.1) can take one of several forms:

- an explicit solution that expresses y as an elementary function of t;

- an explicit solution that expresses y in terms of special functions of t;

- an implicit solution that relates y and t algebraically without expressing y as a function of t;

- a solution that relates y and t through a formula involving integrals.

By *elementary functions* we mean the standard functions of calculus: polynomials, exponentials and logarithms, trigonometric functions and their inverses, and all combinations of these functions through algebraic operations and compositions. By *special functions* we mean various non-elementary functions that mathematicians have given names to, often because they arise as solutions of particularly important differential equations.

As a practical matter, a solution from **DSolve** is most useful when it expresses y explicitly in terms of built-in *Mathematica* functions of t; these include the elementary functions and many special functions. In the following example, we illustrate how to use **DSolve** and its output in this case. See Section 5.4 for examples involving other types of solutions you might get with **DSolve**.

Example 5.1 Consider the linear differential equation

$$\frac{dy}{dt} = t^2 + y.$$

You can find the general solution to this equation in *Mathematica* by typing:

    DSolve[y'[t] == t^2 + y[t], y[t], t]

$$\left\{\left\{y[t] \to -2 - 2t - t^2 + e^t\,C[1]\right\}\right\}$$

The solution of the differential equation is the expression following the arrow. Notice that *Mathematica* produces the answer in terms of an arbitrary constant C[1]. (For higher order equations, there will be as many arbitrary constants as the order of the equation.)

You can obtain specific solutions by choosing specific values for C[1]. In particular, you can find the solution satisfying a given initial condition by imposing the initial condition on the general solution and solving for C[1]. Alternatively, you can specify the initial condition as well as the differential equation when you invoke **DSolve**. To solve the initial value problem

$$\frac{dy}{dt} = t^2 + y, \qquad y(0) = 3,$$

type:

    sol1 = DSolve[{y'[t] == t^2 + y[t], y[0] == 3}, y[t], t]

$$\left\{\left\{y[t] \to -2 + 5\,e^t - 2t - t^2\right\}\right\}$$

In this example, we have given the name **sol1** to the output.

Next, suppose you want to plot the solution or find its value at a particular value of t. To plot the solution on the interval $0 \le t \le 2$, you cannot simply type **Plot[y[t], {t, 0, 2}]**, nor can you type **y[2]** to get the value of the solution at $t = 2$. This is because **DSolve** presents the solution in the form of a *transformation rule*, not a function. To display the value of the solution at $t = 2$, type:

```
y[t] /. First[sol1] /. t → 2
```

$$-10 + 5 e^2$$

To get a numerical value of the solution, reevaluate the previous input line with **//N** at the end:

```
y[t] /. First[sol1] /. t → 2 //N
```

```
26.9453
```

If you are going to evaluate the solution many times, it might be convenient to define a function corresponding to the solution. For the current example, you can first type

```
y1[t_] := -2 + 5Exp[t] - 2t - t^2
```

and then type **Plot[y1[t], {t, 0, 2}]** to plot the solution on the interval $0 \le t \le 2$ or **y1[2]** to evaluate the solution at $t = 2$. Alternatively, you can avoid retyping the solution by defining **y1[t]** as follows:

```
y1[t_] = y[t] /. First[sol1]
```

We often want to study a family of solutions obtained by varying the initial condition. Here is a natural way to do this in *Mathematica*. Begin by solving the differential equation with a generic initial value. For example:

```
sol1a = DSolve[{y'[t] == t^2 + y[t], y[0] == c}, y[t], t]
```

$$\{\{y[t] \to -2 + 2 e^t + c e^t - 2 t - t^2\}\}$$

Mathematica expresses the solution formula in terms of the initial value c. Next, view the right-hand side as a function of both t and c. To define such a function in *Mathematica*, type:

```
y1a[t_, c_] = y[t] /. First[sol1a]
```

Now suppose we want to plot the solution curves with initial values $y(0) = -3, -2, \ldots, 3$ on the interval $0 \le t \le 3$. We can type:

```
Plot[Evaluate[Table[y1a[t, c], {c, -3, 3}]], {t, 0, 3},
    AxesLabel → {x, y}]
```

The result of this command is shown in Figure 5.1. We introduced the **Table** command and the graphics option **AxesLabel** in Chapter 3; and we discussed the **Evaluate** command in Section 8.10 of Chapter 8.

5.2   Existence and Uniqueness

The fundamental existence and uniqueness theorem for differential equations guarantees that every initial condition $y(t_0) = y_0$ leads to a unique solution near t_0, provided that the

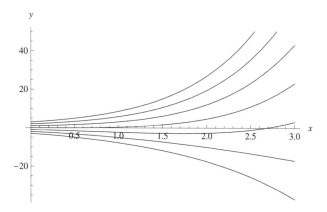

Figure 5.1: Solutions of $dy/dt = t^2 + y$

right-hand side of the differential equation (5.1) is a "nice" function. A theorem like this appears in virtually every textbook on ordinary differential equations.

Theorem 5.1 *Suppose that f and $\partial f/\partial y$ are continuous functions in the rectangle*

$$R = \{(t,y) \ : \ a \leq t \leq b, c \leq y \leq d\},$$

and that the point (t_0, y_0) lies in R. Then the initial value problem

$$dy/dt = f(t,y), \qquad y(t_0) = y_0$$

has a unique solution $y(t)$ that exists at least as long as its graph stays within R.

Notice that by virtue of being a solution, $y(t)$ is differentiable (and hence continuous) for as long as it exists, and that since f is continuous, so is dy/dt.

Graphically, this theorem says that there is a smooth solution curve (or *integral curve*) through every point in R and that the solution curves cannot cross. Thus an initial value problem (IVP) has exactly one solution, but, since there are an infinite number of possible initial conditions, a differential equation has an infinite number of solutions. This principle is implicit in the results obtained above with **DSolve**; when we do not specify an initial condition, the solution depends on an arbitrary constant; when we specify an initial condition, the solution is completely determined.

It is important to remember that the existence and uniqueness theorem only guarantees the existence of a solution *near* the initial point t_0. Consider the initial value problem

$$\frac{dy}{dt} = y^2, \qquad y(0) = 1. \tag{5.2}$$

To solve it symbolically, type:

```
sol2 = DSolve[{y'[t] == y[t]^2, y[0] == 1}, y[t], t]
```

$$\left\{\left\{y[t] \to \frac{1}{1-t}\right\}\right\}$$

To understand where the solution exists, we can graph the expression given by `DSolve` (see Figure 5.2 below).

```
Plot[y[t] /. First[sol2], {t, -1, 2}]
```

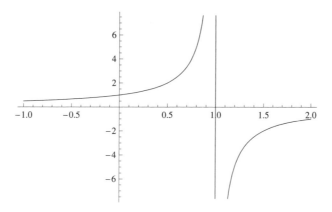

Figure 5.2: Graph of $y = 1/(1-t)$, which for $t < 1$ is the Solution of the IVP (5.2)

We see that the graph has two branches. Since the left branch passes through the initial data point $(0, 1)$, it is the desired solution. Furthermore, we see that the solution exists from $-\infty$ to 1 but does not extend beyond 1. In fact, it becomes unbounded as t approaches 1 from the left. The right branch of the graph depicts more of the function $1/(1-t)$, but it is not part of the solution of the initial value problem.

5.3   Stability of Differential Equations

In addition to existence and uniqueness, the sensitivity of the solution of an initial value problem to the initial condition is a fundamental issue in the theory and application of differential equations. When a differential equation is used to model a physical system, the exact initial condition is generally unknown; instead, there may be a small range of possible initial values. Then in order to assess the amount of uncertainty in predictions made by the model, it is important to know whether solutions that are close to each other at the initial point t_0 remain close together at other values of t. We examine this issue in the following examples.

Example 5.2 Consider the initial value problem

$$\frac{dy}{dt} + 2y = e^{-t}, \qquad y(0) = y_0.$$

We ask: How does the solution $y(t)$ depend on the initial value y_0? More specifically: Does the solution depend continuously on y_0? Do small variations in y_0 lead to small, or large, variations in the solution? Since the solution is

$$y(t) = e^{-t} + (y_0 - 1)e^{-2t},$$

we immediately see that, for any fixed t, the solution $y(t)$ depends continuously on y_0. We can say more. If we let $\tilde{y}(t)$ be the solution of the same equation, but with the initial condition $\tilde{y}(0) = \tilde{y}_0$, then $\tilde{y}(t) = e^{-t} + (\tilde{y}_0 - 1)e^{-2t}$. So, we have

$$|y(t) - \tilde{y}(t)| = |y_0 - \tilde{y}_0|e^{-2t}.$$

Thus for $t \geq 0$ we see that $|y(t) - \tilde{y}(t)|$ is never larger than $|y_0 - \tilde{y}_0|$. In fact, $|y(t) - \tilde{y}(t)|$ decreases as t increases. For $t \leq 0$ the situation is different. When t is a large negative number, the initial difference $|y_0 - \tilde{y}_0|$ is magnified by the large factor e^{-2t}. Even though $y(t)$ depends continuously on y_0, small changes in y_0 lead to large changes in $y(t)$. For example, if $|y_0 - \tilde{y}_0| = 10^{-3}$, then $|y(-7) - \tilde{y}(-7)| = 10^{-3}e^{14} \approx 1203$. These observations are confirmed by Figure 5.3, a plot of the solutions corresponding to initial values $y(0) = 0.97, 1, 1.03$ for $-3 \leq t \leq 3$.

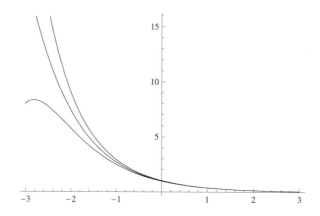

Figure 5.3: Solutions of $dy/dt + 2y = e^{-t}$

Example 5.3 The solution of the initial value problem

$$\frac{dy}{dt} - 2y = -3e^{-t}, \qquad y(0) = y_0$$

is $y(t) = e^{-t} + (y_0 - 1)e^{2t}$. Again, $y(t)$ depends continuously on y_0 for fixed t. Letting $\tilde{y}(t)$ be the solution with initial value $\tilde{y}_0$, we see that

$$|y(t) - \tilde{y}(t)| = |y_0 - \tilde{y}_0|e^{2t}.$$

Now we see that $y(t)$ is very sensitive to changes in the initial value for t large and positive, but insensitive for t negative. These observations are confirmed by Figure 5.4, a plot of the solutions corresponding to initial values $y(0) = 0.97, 1, 1.03$ for $-2 \leq t \leq 2$.

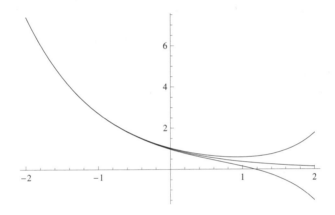

Figure 5.4: Solutions of $dy/dt - 2y = -3e^{-t}$

Often, we are primarily interested in positive values of t, such as when t corresponds to time in a physical problem. If an initial value problem is to predict the future of a physical system effectively, the solution for positive t should be fairly insensitive to the initial value—i.e., small changes in the initial value should lead to small changes in the solution for positive time—for the following reason. As we mentioned at the beginning of this section, for a physical system the initial value y_0 typically is not known exactly. When it is found by measurement, the result is an approximate value $\tilde{y}_0$. Then, if $\tilde{y}(t)$ is the solution corresponding to $\tilde{y}_0$ and the solution is very sensitive to the initial value, $\tilde{y}(t)$ will have little relation to the actual state $y(t)$ of the system as t increases.

An initial value problem whose solution is fairly insensitive to small changes in the initial value as t increases is called *stable*, whereas an initial value problem whose solution is very sensitive to small changes in the initial value is called *unstable*. If all of the solutions of a differential equation are stable (as in Example 5.2), we say the equation is stable, and likewise if all solutions are unstable (as in Example 5.3), we say the equation is unstable. As we have seen, if an equation is unstable and is used over a long time interval, a small error in the initial value can result in a large error later on. Caution should be exercised when using an unstable equation to model a physical problem.

Remark 5.1 We have considered stability as t increases, *i.e.*, stability to the right. We can also consider stability to the left. There are equations that are stable both to the left and to the right, and equations that are unstable both to the left and right.

The following theorem, which we state without proof, is often useful in assessing stability.

Theorem 5.2 *Suppose that $f(t, y)$ has continuous first order partial derivatives in the vertical strip*

$$S = \{(t, y) : t_0 \leq t \leq t_1, -\infty < y < \infty\},$$

and suppose there are numbers K and L such that

$$K \leq \frac{\partial f}{\partial y}(t, y) \leq L, \quad \text{for all } (t, y) \in S.$$

If $y(t)$ and $\tilde{y}(t)$ are solutions of $dy/dt = f(t, y)$ on the interval $t_0 \leq t \leq t_1$ with initial values $y(t_0) = y_0$ and $\tilde{y}(t_0) = \tilde{y}_0$, respectively, then

$$|y_0 - \tilde{y}_0|e^{K(t-t_0)} \leq |y(t) - \tilde{y}(t)| \leq |y_0 - \tilde{y}_0|e^{L(t-t_0)},$$

for all $t_0 \leq t \leq t_1$.

If $L \leq 0$, then the right-hand inequality in the theorem shows that

$$|y(t) - \tilde{y}(t)| \leq |y_0 - \tilde{y}_0|, \quad \text{for all } t_0 \leq t \leq t_1.$$

Thus the solutions differ by no more than the difference in the initial values, and the differential equation is stable (at least for $t_0 \leq t \leq t_1$). Moreover, if $L > 0$, but not too large, and $t_1 - t_0$ is not too large, then

$$|y(t) - \tilde{y}(t)| \leq M|y_0 - \tilde{y}_0|, \quad \text{for all } t_0 \leq t \leq t_1,$$

where $M = e^{L(t_1-t_0)}$ is a moderate-sized constant. Thus the equation is only mildly sensitive to changes in the initial value, and the equation is only mildly unstable. On the other hand, if $K > 0$, then the left-hand inequality in the theorem shows that the solution is sensitive to changes in the initial value, especially over long intervals. We can briefly summarize these results by saying that if $\partial f/\partial y < 0$ in a region of the plane, then solutions in that region get closer together as t increases, while if $\partial f/\partial y > 0$ in a region of the plane, then solutions in that region grow farther apart as t increases. In particular, if $\partial f/\partial y \leq 0$ everywhere, then the differential equation is stable; but if $\partial f/\partial y > 0$ everywhere, then the equation is unstable.

Remark 5.2 The right-hand inequality in the theorem is an example of a *continuous dependence* result; it shows that the solution depends continuously on the initial value.

Let us examine our examples in light of these observations. Rewriting the equation in Example 5.2 as $dy/dt = -2y + e^{-t}$, we see that $f(t, y) = -2y + e^{-t}$ and $\partial f/\partial y = -2$. We can apply the theorem with $t_0 = 0, t_1 = \infty$, and $L = K = -2$ to conclude that

$$|y(t) - \tilde{y}(t)| = |y_0 - \tilde{y}_0|e^{-2t}, \quad \text{for all } t \geq 0,$$

as we found above from the solution formula. Thus the equation is stable. We can also see that the equation is stable just by noting that $\partial f/\partial y < 0$. Similarly, for the equation of Example 5.3, $f(t,y) = 2y - 3e^{-t}$ and $\partial f/\partial y = 2 > 0$, so the equation is unstable.

We can also understand the stability of the differential equations in these two examples by examining the solution formulas. But $\partial f/\partial y$ can be calculated and its sign and size found, even if a solution formula cannot be found. For example, we can immediately tell that the differential equation $dy/dt + t^2 y^3 = \cos t$ is stable because

$$f(t,y) = -t^2 y^3 + \cos t$$

and $\partial f/\partial y = -3t^2 y^2 \leq 0$. Yet neither **DSolve** nor any other standard technique enables us to find a formula solution to this differential equation.

Finally, we note that many equations of the form $dy/dt = f(t,y)$ cannot be classified simply as stable or unstable, because $\partial f/\partial y$ may be negative at some points and positive at others. Nonetheless, we may still be able to determine whether a particular solution is stable (insensitive to its initial value) or unstable (sensitive to its initial value) according to whether $\partial f/\partial y \leq 0$ or $\partial f/\partial y > 0$ along the solution curve. Throughout the book we will see many examples that illustrate the dependence (either sensitive or insensitive) of the solution to an initial value problem on the initial value.

5.4 Different Types of Symbolic Solutions

In Section 5.1, we showed how to evaluate and plot a symbolic solution given by **DSolve** in the ideal case that it outputs an explicit solution. As we mentioned there, many other types of output are possible, and one must deal with them differently. Here we give additional examples illustrating other possibilities.

Example 5.4 Consider the differential equation

$$\frac{dy}{dt} = t + y^2.$$

This equation is neither linear nor separable, and its solutions do not have a formula in terms of elementary functions. However, **DSolve** can still find a formula for the solutions. To get the solution with $y(0) = -3$, type:

```
sol3 = DSolve[{y'[t] == t + y[t]^2, y[0] == -3}, y[t], t]
```

$\left\{\left\{\text{y[t]} \rightarrow \left(-6\,t^{3/2}\,\text{BesselJ}\left[-\frac{2}{3}, \frac{2\,t^{3/2}}{3}\right]\text{Gamma}\left[\frac{1}{3}\right] - 3^{1/3}\,t^{3/2}\,\text{BesselJ}\left[-\frac{4}{3}, \frac{2\,t^{3/2}}{3}\right]\text{Gamma}\left[\frac{2}{3}\right] - \right.\right.\right.$

$3^{1/3}\,\text{BesselJ}\left[-\frac{1}{3}, \frac{2\,t^{3/2}}{3}\right]\text{Gamma}\left[\frac{2}{3}\right] + 3^{1/3}\,t^{3/2}\,\text{BesselJ}\left[\frac{2}{3}, \frac{2\,t^{3/2}}{3}\right]\text{Gamma}\left[\frac{2}{3}\right]\right)\Big/$

$\left(2\,t\left(3\,\text{BesselJ}\left[\frac{1}{3}, \frac{2\,t^{3/2}}{3}\right]\text{Gamma}\left[\frac{1}{3}\right] + 3^{1/3}\,\text{BesselJ}\left[-\frac{1}{3}, \frac{2\,t^{3/2}}{3}\right]\text{Gamma}\left[\frac{2}{3}\right]\right)\right)\Big\}\Big\}$

This formula contains two special functions that are built into *Mathematica*: **BesselJ** and **Gamma**. We can evaluate and plot them exactly as we do with elementary functions like

Exp. Now to understand where the solution **sol3** exists, we graph **sol3** (see Figure 5.5 below).

```
Plot[y[t] /. First[sol3], {t, 0, 6}]
```

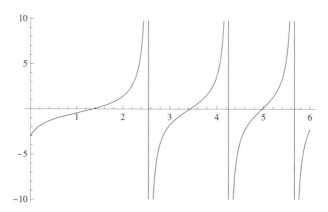

Figure 5.5: The Solution of $dy/dt = t + y^2$, $y(0) = -3$

We see that the graph has several branches. Since the left branch passes through the initial data point $(0, -3)$, it is the desired solution.

Example 5.5 Even when variables can be separated, the solution might be accompanied by a warning message, or it might only be given as an **InverseFunction** object. The differential equation

$$\frac{dy}{dt} = \frac{y^4 + 1}{y^5}$$

is separable, but *Mathematica* gives the solution to this nonlinear differential equation as an **InverseFunction** object:

```
sol4 = DSolve[y'[t] == (y[t]^4 + 1)/y[t]^5, y[t], t]
```

Solve::tdep : The equations appear to involve the variables

to be solved for in an essentially non-algebraic way. $\gg$

$$\left\{\left\{\text{y[t]} \rightarrow \text{InverseFunction}\left[-\frac{1}{2}\text{ArcTan}[\#1^2] + \frac{\#1^2}{2}\&\right][t + \text{C[1]}]\right\}\right\}$$

The `Solve::tdep` message can be ignored; it appears because *Mathematica* cannot find an explicit expression for **y[t]**. *Mathematica* has essentially solved the differential equation; however, it was unable to solve for **y[t]**. In other words, the above answer is *Mathematica*'s convoluted way of giving us an implicit solution. Without an explicit formula for y in terms of t, we must take a different approach to plotting and evaluating solutions.

First we must extract the implicit solution from the *Mathematica* output. In this output, #1 is a dummy variable and the expression $\texttt{InverseFunction}\left[-\frac{1}{2}\texttt{ArcTan[\#1}^2\texttt{]} +\right.$ $\left.\frac{\texttt{\#1}^2}{2}\texttt{\&}\right]\left[\texttt{t}+\texttt{C[1]}\right]$ represents the implicit equation $-\frac{1}{2}\arctan y^2 + \frac{y^2}{2} = t + C[1]$. Now we can extract this equation by solving for $C[1]$ in terms of t and y as follows:

```
sol4eqn = Solve[y == y[t] /. First[sol4], C[1]]
```

$$\left\{\left\{\ \texttt{C[1]} \rightarrow -\texttt{t} + \frac{y^2}{2} - \frac{\texttt{ArcTan}[y^2]}{2}\right\}\right\}$$

The solution curves are the level curves of this expression, which we make into a function as follows:

```
sol4func[t_, y_] = C[1] /. First[sol4eqn]
```

$$-\texttt{t} + \frac{y^2}{2} - \frac{\texttt{ArcTan}[y^2]}{2}$$

Now we use `ContourPlot` to plot several solutions, say for $0 \le t \le 5$ and $0 \le y \le 3$:

```
ContourPlot[sol4func[t, y], {t, 0, 5}, {y, 0, 3},
    FrameLabel → {t, y}, ContourShading → None]
```

The result is shown in Figure 5.6.

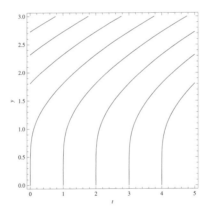

Figure 5.6: Solutions of $dy/dt = (y^4 + 1)/y^5$

Unfortunately, when `DSolve` cannot find an explicit solution, it gives an error message when you try to specify an initial condition. Nonetheless, you can easily find the constant of integration for a given initial condition by plugging the appropriate values of t and y into `sol4func`. So, to plot the solution with initial condition $y(0) = 1$ on the interval $0 \le t \le 5$, type:

```
c = sol4func[1, 0];
```

```
ContourPlot[sol4func[t, y] == c], {t, 0, 5}, {y, 0, 4},
    FrameLabel → {t, y}]
```

The result is shown in Figure 5.7.

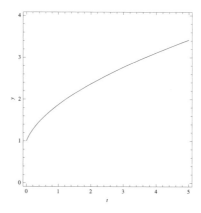

Figure 5.7: Solution of $dy/dt = (y^4 + 1)/y^5$ with $y(0) = 1$

To plot a family of solutions, say with initial conditions $y(0) = 1, 1.2, 1.4, \ldots, 2$, over the same range, type:

```
ContourPlot[sol4func[t,y], {t, 0, 5}, {y, 0, 4},
    Contours → Table[sol4func[0, y0], {y0, 1, 2, 0.2},
    FrameLabel → {t, y}, ContourShading → None]
```

To evaluate a particular solution at a particular value of t, you can substitute the appropriate value of t into **sol4func**, set it equal to the constant of integration you found from the initial condition, and solve for y. One way to do this is as follows:

```
FindRoot[sol4func[y, 4] == c, {y, 3}]
```

$\{\{y \rightarrow 3.11167\}\}$

Here **c** is the value we previously computed for the initial condition $y(0) = 1$, and we have solved for the value of $y(4)$ using *Mathematica*'s **FindRoot** command. We could not use *Mathematica*'s **Solve** command here since, as we saw earlier in this section, *Mathematica* cannot find an explicit expression for y; the transcendental equation **sol4func[y, 4] == c** cannot be solved by algebraic methods. In this case, we used **FindRoot** to solve numerically for y using an initial guess based on the graph in Figure 5.7.

Example 5.6 The differential equation

$$\frac{dy}{dt} = e^{-y} + y$$

is separable, as in the previous example, but this time **DSolve** can't do the y integral:

```
sol5 = DSolve[y'[t] == Exp[-y[t]] + y[t], y[t], t]
```

$$\left\{\left\{\text{y}[\text{t}] \rightarrow \text{InverseFunction}\left[\int_1^{\#1} \frac{e^{K[1]}}{1 + e^{K[1]} K[1]} \, dK[1] \; \& \right][t + C[1]]\right\}\right\}$$

In this output, **K[1]** is a dummy variable, and the expression $\int_1^{\#1} \frac{e^{K[1]}}{1+e^{K[1]}K[1]} \, dK[1]$

represents the definite integral $\int_1^y e^x/(1 + e^x x)dx$. You can use a formula like this in conjunction with a numerical integration routine (such as **NIntegrate**) to obtain a highly accurate approximate solution. However, the numerical methods we will describe in Chapter 7 are generally adequate for such equations.

Remark 5.3 Although we have focused on first order equations in this section, **DSolve** also solves higher order equations (see Chapter 9) and systems of equations (see Chapter 12). The online help for **DSolve** gives examples under **Scope**.

There are still many differential equations that *Mathematica* cannot solve symbolically in any of the forms we have described. This does not mean that there is no solution, just that *Mathematica* cannot find a formula for the solution. As we saw in Section 5.2, solutions to $dy/dt = f(t, y)$ are guaranteed to exist for all reasonably nice functions f. You can use the methods of Chapters 6 and 7 to analyze equations for which no formula solution is available.

Finally, it is important to realize that *Mathematica*, like any software system, can occasionally produce misleading or incorrect results. As the following example shows, a good theoretical understanding of the nature of solutions to differential equations is a valuable guide in interpreting computer-generated results, and therefore in identifying situations in which *Mathematica* has produced a misleading or incorrect result.

Example 5.7 Consider the initial value problem

$$\frac{dy}{dt} = t^2 + y^2, \qquad y(0) = 1.$$

The commands

```
sol6 = DSolve[{y'[t] ==,t^2 + y[t]^2, y[0],==1}, y[t], t];
y6[t_] = y[t] /. First[sol6];
Plot[y6[t], {t, -1, 1}]
```

produce the graph in Figure 5.8. The solution should be continuous, but the graph has a jump at $t = 0$. According to the initial condition, the solution should be 1 at $t = 0$, but typing **y6[0]** gives an error message and the answer **Indeterminate**. The reason is that when *Mathematica* substitutes $t = 0$, its solution formula yields the indeterminate form $0/0$. This would not be such a serious problem if the formula gave correct answers for t near 0, and indeed typing (for instance) **y6[0.001]** gives the answer **1.001**. However, typing **y6[-0.001]** yields **-1.001**. This is not plausible, because for y to change from

-1.001 to 1 as t changes from -0.001 to 0 would require dy/dt to be very large, whereas the differential equation requires that $dy/dt = t^2 + y^2$, which is not large in this range of t and y. We presume that **DSolve** used a solution method that assumed at some point that $t > 0$. Since its solution method solves the differential equation both for $t > 0$ and $t < 0$, the presumptive assumption of positive t yields a discontinuity at $t = 0$.

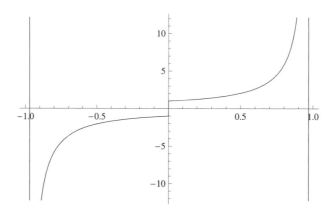

Figure 5.8: The "Solution" of $dy/dt = t^2 + y^2$, $y(0) = 1$, According to **DSolve**

Chapter 6

A Qualitative Approach to Differential Equations

In this chapter, we discuss a *qualitative approach* to the study of differential equations, and obtain qualitative information about the solutions directly from the differential equation, without the use of a solution formula.

Consider the general first order differential equation

$$\frac{dy}{dt} = f(t, y). \tag{6.1}$$

We can obtain qualitative information about the solutions $y(t)$ by viewing (6.1) geometrically. Specifically, we can obtain this information from the direction field of (6.1). Recall that the direction field is obtained by drawing through each point in the (t, y)-plane a short line segment with slope $f(t, y)$. Solutions, or integral curves, of (6.1) have the property that at each of their points they are tangent to the direction field at that point, and therefore the general *qualitative* nature of the solutions can be determined from the direction field. Direction fields can be drawn by hand for some simple differential equations, but *Mathematica* can draw them for any first order equation. We illustrate the qualitative approach with two examples.

6.1   Direction Field for a First Order Linear Equation

Consider the equation

$$\frac{dy}{dt} = e^{-t} - 2y. \tag{6.2}$$

Mathematica's command for plotting direction fields is **VectorFieldPlot**, which can be found in the **VectorFieldPlots** package. To load this package, type:

```
<<VectorFieldPlots`
```

Note the character at the end of the name of the package is a single *backquote*, or *grave* accent. See the *Packages* section of Chapter 3 for important information on using packages. To plot the direction field of (6.2) on the rectangle $-2 \le t \le 3$, $-1 \le y \le 2$, type the following sequence of commands:

```
<< VectorFieldPlots`
VectorFieldPlot[{1, E^(-t) - 2y}, {t, -2, 3}, {y, -1, 2},
    Axes → True, Ticks → False, Frame → True, AspectRatio → 1]
```

The **1** in **{1, E^(-t) - 2y}** is present because **VectorFieldPlot** normally plots the vector field of a system of two differential equations, and we only have a single equation here. (See the *Remarks* at the end of Chapter 13.) The default settings for the options used in the **VectorFieldPlot** command—namely, **Axes**, **Ticks**, **Frame**, and **AspectRatio**—are **False**, **Automatic**, **False**, and **Automatic**, respectively. They are discussed in Section 8.9; see also the online help. The result is displayed in Figure 6.1.

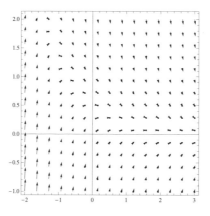

Figure 6.1: Direction Field for Equation (6.2)

While this picture is 'correct,' it is somewhat hard to read because many of the vectors are quite small. One can get a better picture by rescaling the arrows so that they do not vary in magnitude—for example, by dividing each vector $(1, e^{-t} - 2y)$ by its length = $\sqrt{1 + (e^{-t} - 2y)^2}$ so that the resulting vectors all have length 1.

```
VectorFieldPlot[{1, E^(-t) - 2y}/√1 + (E^(-t) - 2y)^2,
    {t, -2, 3}, {y, -1, 2}, Axes → True, Ticks → False,
    Frame → True];
```

We have suppressed the plot because we will use a simpler method to accomplish the same thing—namely, the option **ScaleFunction → (1 &)**, which ensures that the arrows have equal length (see Chapter 13).

```
VectorFieldPlot [{1, E^(-t) - 2y}, {t, -2, 3}, {y, -1, 2},
    ScaleFunction → (1 &), Axes → True, Ticks → False,
    Frame → True];
```

The result is shown in Figure 6.2. The direction field strongly suggests that all solutions approach zero as $t \to \infty$. The general solution of equation (6.2) is $e^{-t} + ce^{-2t}$, so in fact, this is correct.

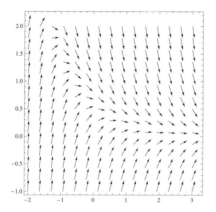

Figure 6.2: Improved Direction Field for Equation (6.2)

Figure 6.3 shows several solution curves superimposed on the direction field. Note from equation (6.2) itself that $y' = 0$ when $2y = e^{-t}$, and thus maximum points on the solution curves in Figure 6.3 occur on the curve $2y = e^{-t}$. In Figure 6.3, the curve $y = \frac{1}{2}e^{-t}$ is indicated with a dashed line.

Exercise 6.1 Pursue this idea further by differentiating (6.2) and showing that the inflection points on the solution curves lie along the curve $y = \frac{3}{4}e^{-t}$.

6.2 Direction Field for a Non-Linear Equation

Equation (6.2) can be solved explicitly because it is linear. Now let us consider an example that cannot be solved explicitly in terms of elementary functions (though **DSolve** does find an explicit solution in terms of Bessel functions):

$$\frac{dy}{dt} = y^2 + t. \tag{6.3}$$

Its direction field is obtained with the *Mathematica* commands:

```
VectorFieldPlot [{1, y^2 + t}, {t, -2, 2}, {y, -2, 2},
    ScaleFunction → (1 &), Axes → True,
```

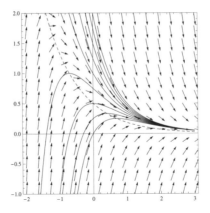

Figure 6.3: Direction Field and Solution Curves for Equation (6.2)

Ticks → False, Frame → True, FrameLabel → {t, y}]

The result is shown in Figure 6.4. From the plot, it appears that all solutions eventually become positive. Figure 6.4 also suggests that all solutions approach infinity. Does this happen at a finite value of t or only as $t \to \infty$? It is impossible to decide on the basis of Figure 6.4, but in fact for each solution $y(t)$ there is a finite value t^* such that $\lim_{t \to t^*} y(t) = \infty$. One can see an indication of this as follows. For $t \geq 1$, the right-hand side of equation (6.3) is $\geq y^2 + 1$. But the equation

$$\frac{dy}{dt} = y^2 + 1 \tag{6.4}$$

is separable, and can be rewritten as

$$\frac{dy}{y^2 + 1} = dt$$

and thus can be solved explicitly; the solution is $\arctan y = t + C$ or $y = \tan(t + C)$. But $\tan(t + C) \to +\infty$ as $t \to \frac{\pi}{2} - C$ (from the left). Since the solutions of (6.4) blow up in finite time, then any solution of (6.3) whose domain of definition includes a value of $t \geq 1$ must blow up in finite time. By the uniqueness of solutions and a glance at the direction field in Figure 6.4, it is clear that *all* solutions must blow up in finite time.

Exercise 6.2 Use *Mathematica* to graph the direction fields for some of the differential equations in your textbook. If you can solve any of the equations explicitly, try superimposing some of the solution curves on top of the direction field.

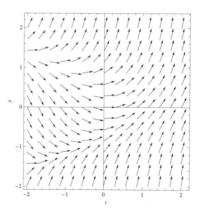

Figure 6.4: Direction Field for Equation (6.3)

6.3   Autonomous Equations

Equations of the form

$$\frac{dy}{dt} = f(y), \tag{6.5}$$

which do not involve t in the right-hand side, are called *autonomous* equations. If a physical system follows rules of evolution that do not change with time, then the evolution of the system is governed by an autonomous equation. Such equations are particularly amenable to qualitative analysis.

Consider equation (6.5). To be concrete, suppose $f(y)$ has two zeros y_1 and y_2, called *critical points* of the differential equation. Furthermore, suppose the graph of $f(y)$ is as shown in Figure 6.5. Then by considering the properties of $f(y)$, the direction field of (6.5) can be easily understood and drawn by hand. At the points (t, y_1) and (t, y_2) the slope is zero; at points (t, y) with $y < y_1$, the slope is positive and increases from 0 to ∞ as y decreases from y_1 to $-\infty$; at points (t, y) with $y_1 < y < y_2$, the slope is negative, and first decreases and then increases as y increases from y_1 to y_2; at points (t, y) with $y > y_2$ the slope is positive and increases from 0 to ∞ as y increases from y_2 to ∞; finally, the slopes along any horizontal line are constant (since f does not depend on t).

The direction field thus has the general appearance shown in Figure 6.6. Note that Figure 6.5 depicts $f(y)$ *vs.* y, whereas Figure 6.6 shows y *vs.* t. Several facts are suggested by the direction field, namely, that: (i) there are two constant solutions $y = y_1$ and $y = y_2$; (ii) solutions starting above y_2 tend to ∞; and (iii) all other solutions tend to y_1 as $t \to \infty$. We can derive these properties of the solutions $y(t)$ directly from the graph of $f(y)$ as follows:

1. The constant functions $y(t) = y_1$ and $y(t) = y_2$ are solutions, called the *equilibrium solutions*. They are the unique solutions that satisfy the initial conditions $y(0) = y_1$ and $y(0) = y_2$, respectively.

f(y)

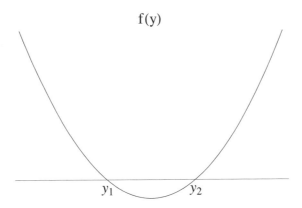

y_1 y_2

Figure 6.5: Right-hand Side of Equation (6.5)

2. Consider a solution $y(t)$ with $y(0) < y_1$. Then, because of the uniqueness of solutions, it must be that $y(t) < y_1$ for all t and

$$y'(t) = f(y(t)) > 0.$$

Hence $y(t)$ is an increasing function. Since it is also bounded above, it has a limit at infinity

$$\lim_{t \to \infty} y(t) = b \le y_1. \tag{6.6}$$

Could it happen that $b < y_1$? The answer is no, because if $b < y_1$, then

$$y'(t) = f(y(t)) \to f(b) > 0. \tag{6.7}$$

But equations (6.6) and (6.7) say that $y(t)$ approaches a horizontal asymptote at the same time that its slope is "permanently" bigger than a positive number. The resulting contradiction guarantees that

$$\lim_{t \to \infty} y(t) = y_1.$$

3. If $y(t)$ is a solution with $y(0) > y_2$, then $y(t) > y_2$ for all t. Also

$$y'(t) = f(y(t)) > 0.$$

Hence $y(t)$ is once again increasing. In fact, $y(t) \to \infty$. This can be shown by an argument similar to that used in part (b). Sometimes, as we shall see from the example below, $y(t)$ actually reaches ∞ in finite time.

4. Now consider $y_1 < y_0 < y_2$. Then if $y(t)$ is a solution with $y(0) = y_0$, we must have $y_1 < y(t) < y_2$ for all t. Also, $y'(t) = f(y(t)) < 0$. Hence $y(t)$ is a decreasing function. As in part 2, it is not difficult to show that $\lim_{t \to \infty} y(t) = y_1$.

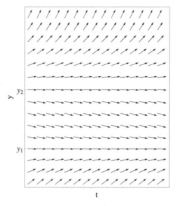

Figure 6.6: Direction Field for Equation (6.5)

Note that in this analysis the properties of $y(t)$ are determined solely from the sign of $f(y)$.

We call the critical point y_1 *asymptotically stable* and the critical point y_2 *unstable*, since solutions that start near y_1 converge to y_1, while those that start near y_2 not only do not converge to y_2, but eventually move away from y_2.

Exercise 6.3 Let $\bar{y}$ be the value between y_1 and y_2 where f takes on its minimum. Show that if $\bar{y} < y(0) < y_2$, then $y(t)$ has an inflection point where $y(t) = \bar{y}$.

6.3.1 Examples of Autonomous Equations

Now we examine a specific example, the equation

$$\frac{dy}{dt} = y^2 - y. \tag{6.8}$$

Its direction field has the general appearance of Figure 6.6, with $y_1 = 0$ and $y_2 = 1$. But we can actually derive an explicit formula solution. If $y \neq 0$ and $y \neq 1$, we can solve by separating variables:

$$
\begin{aligned}
t + C = \int dt = \int \frac{dy}{y^2 - y} \\
= \int \left(\frac{1}{y - 1} - \frac{1}{y} \right) dy \\
= \ln |y - 1| - \ln |y| \\
= \ln \left| \frac{y - 1}{y} \right| \\
= \ln \left| 1 - \frac{1}{y} \right|.
\end{aligned}
$$

Exponentiating both sides gives

$$1 - \frac{1}{y} = ke^t,$$

where $k = \pm e^C$. Solving for y, we find that $y(t) = 1/(1 - ke^t)$. Noting that $y_0 = y(0) = 1/(1 - k)$, we have

$$y(t) = \frac{y_0}{(1 - y_0)e^t + y_0}. \tag{6.9}$$

Although formula (6.9) was derived under the assumption that $y_0 \neq 0, 1$, it is easily seen to be valid for all values of y_0.

Mathematica can solve this equation and plot the solution curves. Here is a sequence of commands that does so.

```
sol = DSolve[{y'[t] == y[t]^2 - y[t], y[0] == c}, y[t], t];
y[t_, c_] := y[t] /. First[sol]
Plot[Evaluate[Table[y2[t, c], {c, -1, 2, 0.25}]], {t, 0, 3},
    Frame → True, FrameLabel → {t, y}, PlotRange → {-7, 8}]
```

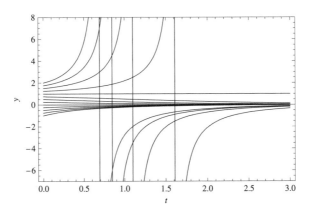

Figure 6.7: Solution Curves for Equation (6.8)

Figure 6.7 contains the actual solution curves for the differential equation—as drawn by Mathematica. Note that the solution curves corresponding to $y(0) > 1$ go to infinity in finite time, and Mathematica includes the vertical asymptotes.

Exercise 6.4

(a) Use formula (6.9) to verify the properties of $y(t)$ obtained above by the qualitative method. Determine where the solutions are increasing or decreasing and whether there are limits as $t \to \infty$.

(b) Suppose $y_0 > 1$. Use formula (6.9) to show that $y(t) \to \infty$ in finite time. Find the time t^* at which this happens.

(c) Show that $y(t) = 0$ and $y(t) = 1$ are the equilibrium solutions. Does formula (6.9) yield these solutions?

Here is another example. The left graph in Figure 6.8 shows a function $f(y)$ with three zeros; thus the autonomous differential equation $y' = f(y)$ has three critical points. Let's look at the critical point in the middle. Solutions starting below b have a positive slope, so they increase toward b. Solutions starting above b have a negative slope, so they decrease toward b. Therefore, we expect b to be an asymptotically stable equilibrium solution. A similar analysis of the signs of $f(y)$ suggests that the other two critical points are unstable. The right graph in Figure 6.8 shows a few of the solution curves. As you can see, this graph illustrates the conclusions that we drew from our analysis of the graph of $f(y)$.

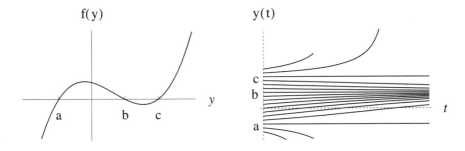

Figure 6.8: An Autonomous Equation with Three Critical Points

Problem Set B

First Order Equations

The solution to Problem 5 appears in the *Sample Solutions*.

1. Consider the initial value problem

$$ty' + 3y = 5t^2, \qquad y(2) = 5.$$

   (a) Solve using **DSolve**. Define the solution function $y(t)$ in *Mathematica*, and then determine its behavior as t approaches 0 from the right and as t becomes large. This can be done by plotting the solution on intervals such as $0.5 \le t \le 5$ and $0.2 \le t \le 20$.

   (b) Change the initial condition to $y(2) = 3$. Determine the behavior of this solution, again by plotting on intervals such as those mentioned in part (a).

   (c) Find a general solution of the differential equation by solving

$$ty' + 3y = 5t^2, \qquad y(2) = c.$$

   Now find the solutions corresponding to the initial conditions

$$y_j(2) = j, \qquad j = 3, \dots, 7.$$

   Plot the functions $y_j(x)$, $j = 3, \dots, 7$, on the same graph. Describe the behavior of these solutions for small positive t and for large t. Find the solution that is not singular at 0. Identify its plot in the graph.

2. Consider the differential equation

$$ty' + 2y = e^t. \tag{B.1}$$

   With initial condition $y(1) = 1$, this has the solution

$$y(t) = \frac{e^t(t-1) + 1}{t^2}.$$

69

(a) Verify this using *Mathematica*, both by direct differentiation and by using `DSolve`.

(b) Graph $y(t)$ on the interval $0 < t < 2$. Describe the behavior of the solution near $t = 0$ and for large values of t.

(c) Plot the solutions $y_j(t)$ of (B.1) corresponding to the initial conditions

$$y_j(1) = j, \qquad j = -3, -2, \ldots, 2, 3,$$

all on the same graph.

(d) What do the solutions have in common near $t = 0$? for large values of t? Is there a solution to the differential equation that has no singularity at $t = 0$? If so, what is it?

3. Consider the initial value problem

$$ty' + y = 2t, \qquad y(1) = c.$$

(a) Solve it using *Mathematica*.

(b) Evaluate the solution with $c = 0.8$ at $t = 0.01, 0.1, 1, 10$. Do the same for the solutions with $c = 1$ and $c = 1.2$.

(c) Plot the solutions with $c = 0.8, 0.9, 1.0, 1.1, 1.2$ together on the interval $(0, 2.5)$.

(d) How do changes in the initial data affect the solution as $t \to \infty$? as $t \to 0^+$?

4. Solve the initial value problem

$$y' - 2y = \sin 2t, \qquad y(0) = c.$$

Use *Mathematica* to graph solutions for $c = -0.5, -0.45, \ldots, -0.05, 0$. Display all the solutions on the same interval between $t = 0$ and an appropriately chosen right endpoint. Explain what happens to the solution curves as t increases. You should identify three distinct types of behavior. Which values of c correspond to which behaviors? Now, based on this problem, and the material in Chapters 5 and 6, discuss what effect small changes in initial data can have on the global behavior of solution curves.

5. Consider the differential equation

$$\frac{dy}{dt} = \frac{t - e^{-t}}{y + e^y}$$

(*cf.* Problem 7, Section 2.2 in Boyce and DiPrima).

(a) Solve it using `DSolve`. Observe that the solution is given implicitly. Express it in the form

$$f(t, y) = c.$$

(b) Use **ContourPlot** (see Section 3.7.6) to examine the solution curves. For your t and y ranges, you might use $\{t, -1, 3\}$ and $\{y, -2, 2\}$. Plot 30 contours.

(c) Plot the solution satisfying the initial condition $y(1.5) = 0.5$.

(d) To find a numerical value for the solution $y(t)$ from part (c) at a particular value of t, you can solve the equation

$$f(t, y) = f(1.5, 0.5)$$

for y. Because there may be multiple solutions, you should look for one near $y = 0.5$. This can be done using **FindRoot** (see Section 3.6). Find $y(0), y(1)$, $y(1.8), y(2.1)$. Mark these values on your plot.

6. Consider the differential equation

$$e^y + (te^y - \sin y)\frac{dy}{dt} = 0.$$

(a) Solve using **DSolve**. Observe that the solution is given implicitly. Express it in the form

$$f(t, y) = c.$$

(b) Use **ContourPlot** (see Section 3.7.6) to examine the solution curves. For your t and y ranges, you might use $\{t, -1, 4\}$ and $\{y, 0, 3\}$. Plot 30 contours.

(c) Plot the solution satisfying the initial condition $y(2) = 1.5$.

(d) Find $y(1), y(1.5), y(3)$. Mark these values on your plot. (See Problem 5, part (d) for suggestions.)

7. Consider the differential equation

$$y' = \frac{t^2}{1 + y^2}$$

(*cf.* Problem 8, Section 2.2 in Boyce and DiPrima).

(a) Solve it using **DSolve**. Observe that in some sense *Mathematica* is "too good" in that it finds three rather complicated explicit solutions. Note that two of them are complex-valued. Select the real one and call it **realsol**. In fact, for this solution it is easier to work with an implicit form, so as in Chapter 5 you can type

```
f[t_,y_] = C[1] /. First[Solve[y == y[t] /. realsol,
    C[1]]]
```

to solve for the constant **C[1]** in terms of t and y and use it to define a function $f(t, y)$. The solution curves are then the level curves of f.

(b) Use **ContourPlot** (see Section 3.7.6) to examine the solution curves. For both your t and y ranges, you might use $\{$**t, -1.5, 1.5**$\}$. Plot 30 contours.

(c) Plot the solution satisfying the initial condition $y(0.5) = 1$.

(d) To find a numerical value for the solution $y(t)$ from part (c) at a particular value of t, you can solve the equation

$$f(t, y) = f(0.5, 1)$$

for y. This can be done using **FindRoot** (see Section 3.6). Find $y(-1), y(0)$, and $y(1)$. Mark these values on your plot.

8. In this problem, we study continuous dependence of solutions on initial data.

(a) Solve the initial value problem

$$y' = y/(1 + t^2), \qquad y(0) = c.$$

(b) Let y_c denote the solution in part (a). Use *Mathematica* to plot the solutions y_c for $c = -10, -9, \ldots, -1, 0, 1, \ldots, 10$ on one graph. Display all the solutions on the interval $-20 \le t \le 20$.

(c) Compute $\lim_{t \to \pm\infty} y_1(t)$.

(d) Now find a constant M such that for all real t, we have

$$|y_a(t) - y_b(t)| \le M|a - b|,$$

for any pair of numbers a and b. Show from the solution formula that $M = 1$ will work if we consider only negative values of t.

(e) Relate the fact in (d) to Theorem 5.2.

9. Use **DSolve** to solve the following differential equations or initial value problems from Boyce and DiPrima. In some cases, *Mathematica* will not be able to solve the equation. (Before moving on to the next equation, make sure you haven't mistyped something.) In other cases, *Mathematica* may give extraneous solutions. (Sometimes these correspond to non-real roots of the equation it solves to get y in terms of t.) If so, you should indicate which solution or solutions are valid. You also might try entering alternative forms of an equation, for example, $M + Ny' = 0$ instead of $y' = -M/N$, or vice versa.

(a) $y' = ry - ky^2$ (Sect. 2.4, Prob. 29),

(b) $y' = t(t^2 + 1)/(4y^3)$, $y(0) = -1/\sqrt{2}$ (Sect. 2.2, Prob. 16),

(c) $(e^t \sin y + 3y) \, dt - (3t - e^t \sin y) \, dy = 0$ (Sect. 2.6, Prob. 8),

(d) $\dfrac{dy}{dt} = (y - 4t)/(t - y)$ (Sect. 2.2, Prob. 30),

(e) $\dfrac{dy}{dt} = (2t + y)/(3 + 3y^2 - t)$, $y(0) = 0$ (Ch. 2, Misc. Prob., Prob. 3).

10. Use **DSolve** to solve the following differential equations or initial value problems from Boyce and DiPrima. See Problem 9 for additional instructions.

(a) $t^3 y' + 4t^2 y = e^{-t}$, $y(-1) = 0$ (Sect. 2.1, Prob. 19),

(b) $y' + (1/t)y = 3 \cos 2t$, $t > 0$ (Sect. 2.1, Prob. 40),

(c) $y' = ty(4 - y)/3$, $y(0) = y_0$, $t > 0$ (Sect. 2.2, Prob. 27),

(d) $(\dfrac{y}{t} + 6t)\, dt + (\ln t - 2)\, dy = 0$, $t > 0$ (Sect. 2.6, Prob. 10),

(e) $y' = \dfrac{t^2 + ty + y^2}{t^2}$ (Sect. 2.2, Prob. 31),

(f) $ty' + ty = 1 - y$, $y(1) = 0$ (Ch. 2, Misc. Prob., Prob. 6).

11. Chapter 6 describes how to plot the direction field for a first order differential equation. For each equation below, plot the direction field on a rectangle large enough (but not too large) to show clearly all of its equilibrium points. Find the equilibria and state whether each is stable or unstable. If you cannot determine the precise value of an equilibrium point from the equation or the direction field, use **FindRoot** or **Solve** as appropriate.

(a) $y' = -y(y - 2)(y - 4)/10$,

(b) $y' = y^2 - 3y + 1$,

(c) $y' = 0.1y - \sin y$.

12. In this problem, we use the direction field capabilities of *Mathematica* to study two nonlinear equations, one autonomous and one non-autonomous.

(a) Plot the direction field for the equation

$$\frac{dy}{dt} = 3 \sin y + y - 2$$

on a rectangle large enough (but not too large) to show all possible limiting behaviors of solutions as $t \to \infty$. Find approximate values for all the equilibria of the system (you should be able to do this with **FindRoot** using guesses based on the direction field picture), and state whether each is stable or unstable.

(b) Plot the direction field for the equation

$$\frac{dy}{dt} = y^2 - ty,$$

again using a rectangle large enough to show the possible limiting behaviors. Identify the unique constant solution. Why is this solution evident from the differential equation? If a solution curve is ever below the constant solution, what must its limiting behavior be as t increases? For solutions lying above the constant solution, describe two possible limiting behaviors as t increases. There is a solution curve that lies along the boundary of the two limiting behaviors. What does it do as t increases? Explain (using the differential equation) why no other limiting behavior is possible.

(c) Confirm your analysis by using **DSolve** on the initial value problem $y' = y^2 - ty$, $y(0) = c$, and then examining different values of c.

13. The solution of the differential equation

$$y' = \frac{2y - t}{2t - y}$$

is given implicitly by $|t - y| = c|t + y|^3$. (This is not what **DSolve** produces, which is a more complicated explicit solution for y in terms of t, but it's what you get by making the substitution $v = y/t$, $y = tv$, $y' = tv' + v$ and separating variables.) However, it is difficult to understand the solutions directly from this algebraic information.

(a) Plot the direction field of the differential equation.

(b) Use **ContourPlot** (see Section 3.7.6) to plot the solutions with initial conditions $y(2) = 1$ and $y(0) = -3$. (Note that the absolute value function is typed **Abs** in *Mathematica*). Use **Show** to put these plots and the vector field plot together on the same graph.

(c) For the two different initial conditions in part (b), use your graphs to estimate the largest interval on which the unique solution function is defined.

14. Consider the differential equation

$$y' = -ty^3.$$

(a) Use *Mathematica* to plot the direction field of the differential equation. Is there a constant solution?

(b) Use **DSolve** to solve the differential equation. Is the constant solution included in formulas given by the output to **DSolve**?

(c) Use **DSolve** to solve the initial value problem

$$y'(t) = -ty^3, \ y(0) = c.$$

(d) Plot the solutions from part (c) on the direction field for initial values $c = -5, -4, \ldots, 5$.

(e) Notice that the solution of the differential equation in part (b) with $C[1] = -1$ is not a solution of the initial value problem in part (c). Plot both solutions from part (b) for $C[1] = -1/4, -1/2, -1$ and superimpose them on the previous graph of the direction field together with the aforementioned solution curves.

15. Consider the critical threshold model for population growth

$$y' = -(2 - y)y.$$

(a) Find the equilibrium solutions of the differential equation. Now draw the direction field, and use it to decide which equilibrium solutions are stable and which are unstable. In particular, what is the limiting behavior of the solution if the initial population is between 0 and 2? greater than 2?

(b) Use **DSolve** to find the solutions with initial values $1.5, 0.3$, and 2.1, and plot each of these solutions. Find the inflection point for the first of these solutions.

(c) Plot the three solutions together with the direction field on the same graph. Do the solutions follow the direction field as you expect them to?

16. Consider the following logistic-with-threshold model for population growth:

$$y' = y(1 - y)(y - 3).$$

(a) Find the equilibrium solutions of the differential equation. Now draw the direction field, and use it to decide which equilibrium solutions are stable and which are unstable. In particular, what is the limiting behavior of the solution if the initial population is between 1 and 3? greater than 3? exactly 1? between 0 and 1?

(b) Next replace the logistic law by the Gompertz model, but retain the threshold feature. The equation becomes

$$y' = y(1 - \ln y)(y - 3).$$

Once again, find the equilibrium solutions and draw the direction field. You will have difficulty "reading the field" between 2.5 and 3. There appears to be a continuum of equilibrium solutions.

(c) Plot the function $f(y) = y(1 - \ln y)(y - 3)$ on the interval $0 \leq y \leq 4$, and then use the **Limit** command to evaluate $\lim_{y \to 0} f(y)$.

(d) Use these plots and the discussion in Chapter 6 to decide which equilibrium solutions are stable and which are unstable. Now use the last plot to explain why the direction field (for $2.5 \leq y \leq 3$) appears so inconclusive regarding the stability of the equilibrium solutions. (*Hint*: The maximum value of f is a relevant number.)

17. This problem is based on Example 1 in Section 2.3 of Boyce and DiPrima: "A tank contains Q_0 lb of salt dissolved in 100 gal of water. Water containing $1/4$ lb of salt per gallon enters the tank at a rate of 3 gal/min, and the well-stirred solution leaves the tank at the same rate. Find an expression for the amount of salt $Q(t)$ in the tank at time t."

The differential equation

$$Q'(t) = 0.75 - 0.03Q(t)$$

models the problem (*cf.* equation (2) in Section 2.3 of Boyce and DiPrima).

(a) Plot the right-hand side of the differential equation as a function of Q, and identify the critical point.

(b) Analyze the long-term behavior of the solution curves by examining the sign of the right-hand side of the differential equation, in a similar fashion to the discussion in Section 6.3.

(c) Use *Mathematica* to plot the direction field of the differential equation. In choosing the rectangle for the direction field be sure to include the point $(0,0)$ and the critical value of Q.

(d) Use the direction field to estimate the limiting amount of salt and to determine how the amount of salt approaches this limit.

(e) Use **DSolve** to find the solution $Q(t)$ and plot it for several specific values of Q_0. Do the solutions behave as indicated in parts (b) and (d)? You should combine the direction field plot from (c) with that of the solution curves.

18. A 10-gallon tank contains a mixture consisting of 1 gallon of water and an undetermined number $S(0)$ of pounds of salt in the solution. Water containing 1 lb/gal of salt begins flowing into the tank at the rate of 2 gal/min. The well-mixed solution flows out at a rate of 1 gal/min. Derive the differential equation for $S(t)$, the number of pounds of salt in the tank after t minutes, that models this physical situation. (*Note*: At time $t = 0$ there is 1 gallon of solution, but the volume increases with time.) Now draw the direction field of the differential equation on the rectangle $0 \le t \le 10$, $0 \le S \le 10$. From your plot:

(a) find the value A of $S(0)$ below which the amount of salt is a constantly increasing function, but above which the amount of salt will temporarily decrease before increasing;

(b) indicate how the nature of the solution function in case $S(0) = 1$ differs from all other solutions.

Now use **DSolve** to solve the differential equation. Reinforce your conclusions above by:

(c) algebraically computing the value of A;

(d) giving the formula for the solution function when $S(0) = 1$;

(e) giving the amount of salt in the tank (in terms of $S(0)$) when it is at the point of overflowing;

(f) computing, for $S(0) > A$, the minimum amount of salt in the tank, and the time it occurs;

(g) explaining what principle guarantees the truth of the following statement: If two solutions S_1, S_2 correspond to initial data $S_1(0)$, $S_2(0)$ with $S_1(0) < S_2(0)$, then for any $t \geq 0$, it must be that $S_1(t) < S_2(t)$.

19. In this problem, we use **DSolve** and **Solve** to model some population data. The procedure will be:

   (i) Assume a model differential equation involving unknown parameters.

   (ii) Use **DSolve** to solve the differential equation in terms of the parameters.

   (iii) Use **Solve** to find the values of the parameters that fit the given data.

   (iv) Make predictions based on the results of the previous steps.

   (a) Let's use the model
   $$\frac{dp}{dt} = ap + b, \quad p(0) = c,$$
   where p represents the population at time t. Check to see that **DSolve** can solve this initial value problem in terms of the unknown constants a, b, c. Then define a function that expresses the solution at time t in terms of a, b, c, and t. Give physical interpretations to the constants a, b and c.

   (b) Next, let's try to model the population of Nevada, which was one of the fastest growing states in the U.S. during the second half of the twentieth century. Here is a table of census data:

Year	Population in thousands
1950	160.1
1960	285.3
1970	488.7
1980	800.5
1990	1201.8
2000	1998.3

   We would like to find the values of a, b, and c that fit the data. However, with three unknown constants we will not be able to fit six data points. Use **Solve** to find the values of a, b, and c that give the correct population for the years 1960, 1970, and 1980. We will later use the data from 1950, 1990, and 2000 to check the accuracy of the model. *Important*: In this part let t represent the time in years since 1960, because *Mathematica* may get stuck if you ask it to fit the data at such high values of t as 1960–1980.

   (c) Now define a function of t that expresses the predicted population in year t using the values of a, b, c found in part (b). Find the population this model gives for 1950, 1990, and 2000, and compare with the values in the table above. Use the model to predict the population of Nevada in the year 2010, and to predict when the population will reach 5 million. How would you adjust these predictions based on the 1950, 1990, and 2000 data? What adjustment to a

and/or b might you make? Finally, graph the population function that the model gives from 1950 to 2050, and describe the predicted future of the population of Nevada, including the limiting population (if any) as $t \to \infty$.

20. Consider $y' = (\alpha - 1)y - y^3$.

   (a) Use **Solve** to find the roots of $(\alpha - 1)y - y^3$. Explain why $y = 0$ is the only real root when $\alpha \leq 1$, and why there are three distinct real roots when $\alpha > 1$.

   (b) For $\alpha = -2, -1, 0$, draw a direction field for the differential equation and deduce that there is only one equilibrium solution. What is it? Is it stable?

   (c) Do the same for $\alpha = 1$.

   (d) For $\alpha = 1.5, 2$, draw the direction field. Identify all equilibrium solutions, and describe their stability.

   (e) Explain the following statement: "As α increases through 1, the stable solution $x = 0$ *bifurcates* into two stable solutions."

21. In Chapter 6, we discussed equation (6.3) and the fact that its solutions blow up in finite time. In this problem, we explore the solutions in a little more detail.

   (a) Use **DSolve** to solve the initial value problem $y' = y^2 + t$ with the initial condition $y(1) = 1$. You might find that *Mathematica* writes the solution in terms of the built-in function **BesselJ**, which was mentioned in Chapter 5.

   (b) Plot the solution you have obtained over the interval $0.75 \leq t \leq 2$. At what value t^* of t does it seem that the solution "blows up"? What happens to your graph past this value of t^*? Do you think it has any validity? Why or why not?

   (c) Compute an upper bound for t^* by using the analysis of Section 6.2, that is, by comparison with the equation $y' = y^2 + 1$ for $t \geq 1$. Compare the graphs of the solution to the given IVP to that of the "approximate" IVP to see that you indeed have an upper bound for t^*.

   (d) Superimpose your graph of the solution to (6.3) on top of the direction field for the equation, and visually verify the tangency of the solution curve to the direction field.

Chapter 7

Numerical Methods

In this book you have seen that many differential equations can be solved explicitly in terms of the functions of calculus. We have seen in Chapter 5 that **DSolve** solves many equations in terms of these functions.

But many other differential equations cannot be solved explicitly in terms of the functions of calculus. Consider, for example, the equation

$$\frac{dy}{dt} = e^{-t^2}.$$

Its solutions are the integrals, or antiderivatives, of e^{-t^2},

$$y(t) = \int e^{-t^2} \, dt + C,$$

but it is known that these integrals cannot be expressed in terms of elementary functions. (*Mathematica* does write this integral in closed form in terms of a function **Erf**, but this function, called the "error function," is *defined* by the formula $\mathrm{erf}(z) = \frac{2}{\sqrt{\pi}} \int_0^z e^{-t^2} \, dt$, and cannot be rewritten in terms of polynomials, exponentials, logs, trigonometric functions, *etc.* This is the simplest example of a *special function* as discussed in the beginning of Chapter 5.)

We will refer to solutions that can be explicitly written in terms of elementary functions or special functions as *formula* or *symbolic solutions*—whether they were obtained via hand-calculation or via **DSolve**. When we cannot solve a differential equation in this way, or if the formula we find is too complicated, we turn to numerical methods to solve an initial value problem. This is similar to a situation in calculus: if we cannot find an antiderivative in terms of elementary functions, we turn to a numerical method such as the trapezoidal rule or Simpson's rule to evaluate a definite integral.

With numerical methods as with qualitative methods, we obtain information about the solution—quantitative in the one case and qualitative in the other—without the use of a solution formula.

7.1   Numerical Solutions Using *Mathematica*

Suppose we are interested in finding the solution to the initial value problem

$$\frac{dy}{dt} = f(t, y), \qquad y(t_0) = y_0$$

on an interval $a \leq t \leq b$ containing t_0, and suppose that we do not have a formula for $y(t)$. In such a situation, our strategy will be to produce a function $y_a(t)$ that both is a good approximation to $y(t)$ and can be calculated for any $t \in [a, b]$. The subscript a on y_a stands for *approximate solution*. Such a function $y_a(t)$ can be found with **NDSolve**, *Mathematica*'s primary numerical differential equation solver.

We illustrate the use of **NDSolve** by considering the initial value problem

$$\frac{dy}{dt} = \frac{t}{y}, \qquad y(0) = 1. \tag{7.1}$$

Its exact solution is

$$y(t) = \sqrt{t^2 + 1}.$$

Since we have an explicit formula for this solution, we will be able to compare the approximate solution $y_a(t)$ and the exact solution $y(t)$.

To obtain $y_a(t)$ on the interval $[0, 2]$, type

```
ivp1 = {y'[t] == t/y[t],y[0] == 1};
sol1 = NDSolve[ivp1,y[t],{t,0,2}]
```

The first argument to **NDSolve** is a list that specifies the differential equation and the initial condition. The second argument specifies the dependent variable **y[t]**. The third argument, **{t, 0, 2}**, is a list consisting of the independent variable and the endpoints of the interval on which we wish to calculate the approximate solution.

The preceding commands produce the output

```
{{ y[t] → InterpolatingFunction[{{0.,2.}}, <>][t]}}
```

This is not quite the desired form of the approximate solution. To get the desired form, we use the replacement operator **/.** and the **First** command (exactly as when we introduced **DSolve** in Chapter 5). We type

```
ya1[t_] = y[t] /. First[sol1]
```

which yields the output

```
InterpolatingFunction[{{0.,2.}}, <>][t]
```

Now the approximate solution **ya1[t]** can be evaluated for any particular $t \in [0, 2]$, and hence can be graphed on the interval $0 \leq t \leq 2$. For example, typing **ya1[1.5]** yields the output **1.80278**. Note that the exact value is $y(1.5) = \sqrt{13}/2 = 1.802775637\ldots$, so that **ya1[1.5]** is an approximate value with six digits of precision.

The command

```
TableForm[Table[{t,ya1[t], √t² +1 },{t,0,2,0.2}],
```

```
TableHeadings → {None, {"t", "ya1[t]", "y[t]"}},
TableSpacing → {1, 2}]
```

produces a table of numerical solution values and exact solution values at the points $0, 0.2,$ $\ldots, 1.8, 2$. The result is shown in Table 7.1. The approximate solution is correct to at

t	ya1[t]	y[t]
0.	1.	1.
0.2	1.0198	1.0198
0.4	1.07703	1.07703
0.6	1.16619	1.16619
0.8	1.28062	1.28062
1.	1.41421	1.41421
1.2	1.56205	1.56205
1.4	1.72046	1.72047
1.6	1.8868	1.8868
1.8	2.05913	2.05913
2.	2.23607	2.23607

Table 7.1: Approximate and Exact Solutions of $dy/dt = t/y, y(0) = 1$

least five digits in all cases. The **NDSolve** command attempts to produce an approximate solution with error less than or equal to 10^{-6}, though sometimes the accuracy may be better or worse. See Sections 7.5 and 7.6 for additional information on accuracy.

A graph of **ya1[t]** can be obtained with the command **Plot[ya1[t], {t, 0, 2}]**. The result is shown in Figure 7.1.

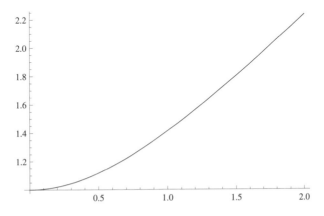

Figure 7.1: Numerical Solution of $dy/dt = t/y, y(0) = 1$

Exercise 7.1 Display both $y_a(t)$ and the exact solution $y(t) = \sqrt{t^2 + 1}$ on the same graph. Can you distinguish between the two curves?

You can also plot a family of approximate solutions with various initial conditions $y(0) = a$. However, **NDSolve** must have a numerical value for the initial condition before it is evaluated. We will circumvent this problem by using delayed evaluation (**:=**). For example, to plot the family of solutions of $dy/dt = t/y$ with initial values $y(0) = 1, 1.2, 1.4, \ldots, 3$ on the interval $[-1, 2]$, type

```
ivp2 = {y'[t] == t/y[t], y[0] == a};
sol2 := NDSolve[ivp2, y[t], {t, -1, 2}]
ya2[t_, a_] := y[t] /. First[sol2]
Plot[Evaluate[Table[ya2[t, a], {a, 1, 3, 0.2}]],
    {t, -1, 2}, AxesOrigin → {0, 0}]
```

The result is shown in Figure 7.2.

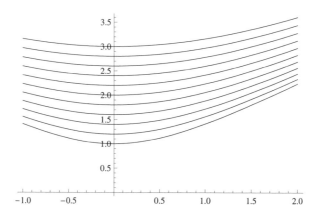

Figure 7.2: A Family of Numerical Solutions

Notice when using **NDSolve** to calculate an approximate solution on the interval $a \le t \le b$, the interval does not have to be of the form $t_0 \le t \le b$; in particular; a can be less than t_0. Also note the option **AxesOrigin → {0,0}** places the origin in Figure 7.2 at the point $(0, 0)$; the origin in Figure 7.1 is at $(0, 1)$.

We have illustrated **NDSolve** with the initial value problem (7.1), which can easily be solved explicitly. We can just as readily apply **NDSolve** to any first order initial value problem. Consider, for example,

$$\frac{dy}{dt} = t + y^4, \qquad y(0) = 1, \tag{7.2}$$

which cannot be solved by **DSolve**. The commands

```
ivp3 = {y'[t] == t+y[t]^4, y[0] == 1};
sol3 := NDSolve[ivp3, y[t], {t, 0, 1}]
ya3[t_] = y[t] /. First[sol3]
```

will produce an approximate solution.

Exercise 7.2 The solution $y(t)$ to (7.2) is increasing for t positive, and there is a finite value t^* such that $\lim_{t \to t^*} y(t) = \infty$; the interval of existence of the solution is $(-\infty, t^*)$. Execute the previous three commands. Does **NDSolve** reveal the value of t^*?

Remark 7.1 Although we have focused on first order equations in this chapter and in Chapter 5, both **NDSolve** and **DSolve** work on higher order equations and systems of equations. See Chapter 8, Chapter 12, and the Sample Solutions for examples.

7.2   Some Numerical Methods

In order to give some idea about how **NDSolve** calculates approximate values, we are going to discuss several numerical methods. We will apply each method to problem (7.1).
   We want to approximate the solution of the initial value problem

$$\frac{dy}{dt} = f(t, y), \qquad y(t_0) = y_0$$

on an interval $t_0 \le t \le b$. For n a positive integer, we divide the interval into n parts using points

$$t_0 < t_1 < t_2 < \cdots < t_n = b.$$

For simplicity, we assume that each part has the same width, or *step size*, $h = t_{i+1} - t_i = (b - t_0)/n$; therefore $t_i = t_0 + ih$. At each point t_i we seek an approximation, which we call y_i, to the true solution $y(t_i)$ at t_i:

$$y(t_i) \approx y_i.$$

7.2.1   The Euler Method

The simplest numerical solution method is due to Euler and is based on the tangent line approximation to a function. Given the initial value y_0, we define y_i recursively by

$$y_{i+1} = y_i + h\,f(t_i, y_i), \qquad i = 0, 1, \ldots, n-1.$$

This formula is derived as follows:

$$
\begin{aligned}
y(t_{i+1}) &\approx y(t_i) + hy'(t_i), & \text{by the tangent line approximation} \\
&= y(t_i) + hf(t_i, y(t_i)), & \text{using the differential equation} \\
&\approx y_i + hf(t_i, y_i), & \text{since } y(t_i) \approx y_i \\
&= y_{i+1}.
\end{aligned}
$$

The approximations $y(t_i) \approx y_i$ should become better and better as h is taken smaller and smaller.

Example 7.1 Consider the initial value problem (7.1):

$$\frac{dy}{dt} = \frac{t}{y}, \qquad y(0) = 1.$$

We wish to approximate $y(0.3)$ using the Euler Method with step size $h = 0.1$ and three steps. We find

$$
\begin{aligned}
t_0 &= 0, \quad t_1 = 0.1, \quad t_2 = 0.2, \quad t_3 = 0.3 \\
y_0 &= 1 \\
y_1 &= y_0 + hf(t_0, y_0) = y_0 + ht_0/y_0 = 1 \\
y_2 &= y_1 + ht_1/y_1 = 1.01 \\
y_3 &= 1.0298.
\end{aligned}
$$

Since $y(0.3) = \sqrt{1.09} = 1.044030...$, we find that

$$\text{Error} = |y(0.3) - y_3| = 0.0142....$$

Next use $h = 0.05$ and 6 steps:

$$
\begin{aligned}
y_0 &= 1 \\
y_1 &= 1 \\
y_2 &= 1.0025 \\
y_3 &= 1.0075 \\
y_4 &= 1.0149 \\
y_5 &= 1.0248 \\
y_6 &= 1.0370
\end{aligned}
$$

$$\text{Error} = |y(0.3) - y_6| = 0.0070 \ldots .$$

If the initial value problem is "sufficiently smooth", then for the Euler Method one can show that the error in stepping from t_0 to any t_j in the interval $t_0 \le t \le b$ satisfies

$$\text{Error} \le Ch,$$

where C is a constant that depends on $f(t, y)$, its partial derivatives, the initial condition, and the interval, but not on h. Moreover, it can be shown that the error is actually proportional to h. Because of this, the Euler Method is called a *first order method*. Note that in the example, cutting the step size in half had the effect of cutting the error approximately in half, as expected for a first order method.

It is also useful to know the *local error*. Suppose $u(t)$ is the solution of the differential equation satisfying $u(t_j) = y_j$. Then the local error in stepping from t_j to t_{j+1} is defined to be

$$e_{j+1} = u(t_{j+1}) - y_{j+1};$$

i.e., the error made in one step assuming the solution value at t_j is y_j. Using Taylor's formula it is easily shown that, for the Euler Method,

$$e_{j+1} = \frac{1}{2}u''(\bar{t}_j)h^2,$$

for some $\bar{t}_j \in (t_j, t_{j+1})$. Thus the local error is proportional to h^2. Because local error provides a simple comparison of methods and is used in the design of numerical solution software, we will state the local error for each method we discuss. The error discussed in the previous paragraph is called *global error*, in order to distinguish it from local error.

The relation between local and global error can be understood intuitively in the following way. We begin by noting that

Global Error at $t_{j+1} = y(t_{j+1}) - y_{j+1} = (y(t_{j+1}) - u(t_{j+1})) + (u(t_{j+1}) - y_{j+1}).$

The second term on the right-hand side of this equation is the local error, which is proportional to h^2. The first term is the difference at t_{j+1} of the two solutions of the differential equation with values $y(t_j)$ and y_j at t_j. The size of this term depends on the size of $y(t_j) - y_j$ and on the differential equation, specifically on whether the difference between $y(t_j)$ and y_j (which serve as initial values at t_j) leads to a smaller or larger difference in the values of the solutions at t_{j+1}, *i.e.*, on whether the differential equation is stable or unstable. These terms were defined in Chapter 5.

Suppose the differential equation is stable. Then its solutions are fairly insensitive to initial values, so that $y(t_{j+1}) - y_{j+1}$ is about the same size as or smaller than $y(t_j) - y_j$. In this situation we see that the accumulated, global error is approximately the sum of the local errors. To be specific, in stepping from t_0 to t_j, we make j local errors, each of which is proportional to h^2. Since $j \le n$, we thus have an accumulated error that is no greater than a constant times

$$nh^2 = \frac{b - t_0}{h}h^2 = (b - t_0)h;$$

i.e., the global error is bounded by a constant times h.

Next, suppose the differential equation is unstable, so that $u(t_{j+1}) - y_{j+1}$ may be larger than $y(t_j) - y_j$. Then the global error at t_{j+1} is the sum of this larger quantity and the local error at t_{j+1}. Thus, in this situation, the global error depends on the stability of the differential equation as well as on the numerical method, and cannot be completely assessed by the simple calculation sketched above. It can still be proved that, as stated earlier, the global error is proportional to h. In this result, stability influences the constant of proportionality: if the differential equation is stable, the constant is of moderate size; if unstable, the constant may be large. The effect of stability and instability on numerical solutions will be discussed further in Section 7.6.

The Euler Method in *Mathematica*.   The Euler Method can be implemented using *Mathematica* as follows:

```
EulerMethod[f_,{t0_,y0_},h_,n_] :=
    (EMStep[{t_,y_}] := N[{t+h,y+h*f[t,y]}];
    NestList[EMStep,{t0,y0},n])
```

To apply this program to the initial value problem (7.1), we first define the function $f(t, y) = t/y$ and then run the routine. So,

```
f[t_, y_] := t/y
em = EulerMethod[f,{0,1},0.1,20]
```

will produce a list of approximate solution values at $t = 0, 0.1, 0.2, \ldots, 2$. These can be displayed using the command **TableForm[em]**. Note that **EulerMethod** begins by defining a new function, called **EMStep**, that performs a single Euler Method step. It then uses **NestList** to iterate **EMStep**.

To approximate the solution at a point t that is not a t_i, we have to *interpolate*. A simple way to do this is to "connect the dots" by drawing straight line segments from each point (t_i, y_i) to the next point (t_{i+1}, y_{i+1}); thus $y_a(t)$ would be the piecewise linear function connecting the computed points (t_i, y_i). To graph $y_a(t)$, type **ListPlot[em, PlotJoined → True]**.

Using the Euler Method to produce an "interpolating function" approximating the solution of an initial value problem is already pre-programmed into *Mathematica* using the **Method→ "ExplicitEuler"** option to **NDSolve**. Thus the problem above could also be solved by typing:

```
Eulerapprox = y[t] /. First[NDSolve[ivp1, y[t], {t, 0, 2},
    StartingStepSize → 1/10, Method → "ExplicitEuler"]];
Show[Plot[Eulerapprox, {t, 0, 2}],
    ListPlot[Table[{t, Eulerapprox}, {t, 0, 2, 0.1}]]]
```

This program produces the same values at the points t_i as produced by **EulerMethod** above, but the plot in Figure 7.3 uses interpolation by higher order polynomials in order to get a smoother graph.

7.2.2   The Improved Euler Method

We again start with the tangent line approximation, replacing the slope $y'(t_i)$ by the average of the two slopes $y'(t_i)$ and $y'(t_{i+1})$. This yields

$$
\begin{aligned}
y(t_{i+1}) &\approx y(t_i) + h \, \frac{y'(t_i) + y'(t_{i+1})}{2} \\
&= y(t_i) + h \, \frac{f(t_i, y(t_i)) + f(t_{i+1}, y(t_{i+1}))}{2}.
\end{aligned}
\tag{7.3}
$$

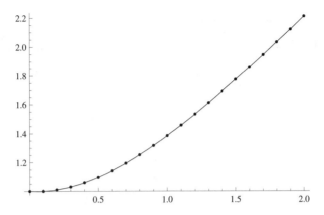

Figure 7.3: An Euler Method Solution

We now use the approximation $y(t_i) \approx y_i$ and the previous Euler approximation $y(t_{i+1}) \approx y_i + hf(t_i, y_i)$ to get

$$
\begin{aligned}
y(t_{i+1}) &\approx y_i + h \,\frac{f(t_i, y_i) + f(t_{i+1}, y_i + hf(t_i, y_i))}{2} \\
&= y_i + h \,\frac{y_i' + f(t_{i+1}, y_i + hy_i')}{2} \\
&= y_{i+1},
\end{aligned}
$$

where $y_i' = f(t_i, y_i)$. Given the initial value y_0, this analysis leads to the recursion formula

$$
y_{i+1} = y_i + h \,\frac{y_i' + f(t_{i+1}, y_i + hy_i')}{2}, \qquad i = 0, 1, \ldots, n - 1.
$$

Example 7.2 We again consider problem (7.1) and find an approximation to $y(0.3)$ with $h = 0.1$. We obtain

$$
\begin{aligned}
y_1 &= y_0 + h \,\frac{y_0' + f(t_1, y_0 + hy_0')}{2} \\
&= y_0 + \frac{h}{2} \left(\frac{t_0}{y_0} + \frac{t_1}{y_0 + ht_0/y_0} \right) \\
&= 1.005 \\
y_2 &= 1.019828 \\
y_3 &= 1.044064.
\end{aligned}
$$

So Error $= |y(0.3) - y_3| = 0.000033\ldots$ Note that the error in the Improved Euler Method with $h = 0.1$ is considerably less than the error in the Euler Method with $h = 0.05$.

The local error for the Improved Euler Method is proportional to h^3.

7.2.3   The Runge-Kutta Method

Given the initial value y_0, the fourth order Runge-Kutta Method is defined recursively by

$$y_{i+1} = y_i + \frac{h}{6}\left(k_1 + 2k_2 + 2k_3 + k_4\right),$$

where

$$
\begin{aligned}
k_1 &= f(t_i, y_i) \\
k_2 &= f\left(t_i + \frac{h}{2}, y_i + \frac{h}{2}k_1\right) \\
k_3 &= f\left(t_i + \frac{h}{2}, y_i + \frac{h}{2}k_2\right) \\
k_4 &= f(t_{i+1}, y_i + hk_3).
\end{aligned}
$$

Note that a weighted average of the "slopes" k_1, k_2, k_3, k_4 is used in the formula for y_{i+1}.

Example 7.3 Consider problem (7.1) and again find an approximation to $y(0.3)$, but now let $h = 0.3$ and take just one step. Then

$$
\begin{aligned}
k_1 &= f(0,1) = 0 \\
k_2 &= f(0 + 0.15, 1 + 0.15(0)) = 0.15 \\
k_3 &= f(0.15, 1 + 0.15(0.15)) = 0.146699 \\
k_4 &= f(0.3, 1 + 0.3(0.146699)) = 0.287354 \\
y_1 &= 1 + \frac{0.3}{6}\left(0 + 2(0.15 + 0.146699) + 0.287354\right) \\
&= 1.044038 \approx y(0.3).
\end{aligned}
$$

So Error $= |y(0.3) - y_1| = 0.000007\ldots$. We see that the error in the Runge-Kutta Method with $h = 0.3$ is much less than the error in the Euler Method with $h = 0.05$, or in the Improved Euler Method, with $h = 0.1$. The local error for the Runge-Kutta Method is proportional to h^5.

Remark 7.2 The methods we have discussed can be applied to any first order equation. Furthermore, they can be applied using different step sizes at each step.

7.2.4   The Adams Methods

We discuss two methods from the Adams family, one an Adams-Bashforth Method and the other an Adams-Moulton Method. The third order Adam-Bashforth Method is defined by the formula

$$y_{i+1} = y_i + \frac{h}{12}(5f(x_{i-2}, y_{i-2}) - 16f(x_{i-1}, y_{i-1}) + 23f(x_i, y_i)). \tag{7.4}$$

One needs to know y_{i-2}, y_{i-1}, and y_i in order to use the formula to determine y_{i+1}. Because of this feature, the method is called a *multistep method*, specifically, a *three-step method*. The previous methods (Euler, Improved Euler, and Runge-Kutta) were *one-step methods*. To use a multi-step method, one must calculate a few initial values using some other (one-step) method. The local error for the third order Adams-Bashforth Method is proportional to h^4.

The fourth order Adams-Moulton Method is

$$y_{i+1} = y_i + \frac{h}{24}(f(x_{i-2}, y_{i-2}) - 5f(x_{i-1}, y_{i-1}) + 19f(x_i, y_i) + 9f(x_{i+1}, y_{i+1})). \quad (7.5)$$

This method is also a three-step method. Since y_{i+1} appears on both sides of the equation, one has to solve the equation for y_{i+1}. Because of this feature, the method is called *implicit*. With the previous methods, the formulas are already solved explicitly for y_{i+1}, and the methods are referred to as *explicit*. The local error for the fourth order Adams-Moulton Method is proportional to h^5.

There are Adams-Bashforth Methods and Adams-Moulton Methods of every order.

The solution y_{i+1} of (7.5), which generally is nonlinear, can be approximated as follows: Let $y_{i+1}^{(0)}$ be an initial approximation; we could, *e.g.*, take $y_{i+1}^{(0)}$ to be defined by the Adams-Bashforth formula (7.4). Then define $y_{i+1}^{(1)}$ in terms of $y_{i+1}^{(0)}$ and y_{i-2}, y_{i-1}, and y_i by

$$y_{i+1}^{(1)} = y_i + \frac{h}{24}(f(x_{i-2}, y_{i-2}) - 5f(x_{i-1}, y_{i-1}) + 19f(x_i, y_i) + 9f(x_{i+1}, y_{i+1}^{(0)})). \quad (7.6)$$

Formula (7.6) is based on the right-hand side of (7.5). Continuing, we can calculate $y_{i+1}^{(2)}$ from $y_{i+1}^{(1)}$ using a formula similar to (7.6), $y_{i+1}^{(3)}$ from $y_{i+1}^{(2)}$, and so on. Under appropriate assumptions, the sequence $y_{i+1}^{(0)}, y_{i+1}^{(1)}, y_{i+1}^{(2)}, \cdots$ converges to y_{i+1}. We thus consider $y_{i+1}^{(1)}, y_{i+1}^{(2)}, \cdots$ to be "improved" approximations to y_{i+1}. Often $y_{i+1}^{(1)}$ is a sufficiently accurate approximation. The process we have outlined for determining $y_{i+1}^{(0)} y_{i+1}^{(1)}, y_{i+1}^{(2)}, \cdots$, is called *fixed-point iteration*.

When an Adam-Bashforth formula and an Adams-Moulton formula are used in this way, we speak of an Adams *predictor-corrector pair*. Adams-Bashforth and Adam Moulton formulas of other orders can be paired in this way.

7.2.5   Backwards Differentiation Formulas

The fourth order Backwards Differentiation Formula is

$$y_{i+1} = \frac{1}{25}(-3y_{i-3} + 16y_{i-2} - 36y_{i-1} + 48y_i + 12hf(x_{i+1}, y_{i+1})).$$

This is an implicit, four-step method. Its local error is proportional to h^5. There are Backwards Differentiation Formulas of every order. Backwards Differentiation Formulas are sometimes referred to as Gear Formulas.

7.3   Inside `NDSolve`

Software for the numerical solution of ordinary differential equations can be based on the methods we have presented, as well as on other methods. By default, **NDSolve** uses the Adams Predictor-Corrector Formulas and the Backwards Differentiation Formulas. It varies both the step size and the order of the methods or formulas, choosing them at each step to try to ensure that the desired accuracy is attained. All efficient modern numerical ODE solvers use variable step size methods, and many use variable order methods.

In addition to choosing step sizes and orders, **NDSolve** chooses between the Adams Formulas and the Backwards Differentiation Formulas, depending on the problem. Initial value problems can be classified as *stiff* or *non-stiff*. **NDSolve** attempts to test for stiffness; it then uses the Adams Formulas for non-stiff problems and the Backwards Differentiation Formulas for stiff problems. Stiff problems are generally more difficult to solve numerically, and the Backwards Differentiation Formulas are particularly effective for such problems. For additional information on numerical methods and software, we refer to L. Shampine, *Numerical Solution of Ordinary Differential Equations*, Chapman & Hall, 1994 and D. Kahaner, C. Moler, and S. Nash, *Numerical Methods and Software*, Prentice Hall, Inc., 1989.

The solver **NDSolve** produces approximate solution values at the sequence of points determined by its selection of step sizes; we refer to these points as internal time points. Most values of t will not coincide with the internal time points, and interpolation must be used to get approximate solution values at the points where you evaluate the approximate solution. The interpolation procedure connects the approximate solution values at the internal time points, not with straight lines (linear polynomials), as mentioned before in the case of the Euler Method, but with higher degree polynomials. The solver **NDSolve** attempts to perform both processes—the calculation of the approximate solution values at the internal time points and the interpolation process—in such a way that the total error in both processes, $|y(t) - y_a(t)|$, does not exceed the desired level, by default 10^{-6}.

You can use the **Method** option with **NDSolve** to select other methods. The setting **Method** $\rightarrow$ **Adams** specifies a method based on the Adams-Moulton Methods. **Method** $\rightarrow$ **BDF** specifies a method based on the Backwards Differentiation Formulas. **Method** $\rightarrow$ **RungeKutta** leads to the Runge-Kutta-Fehlberg Method, which is based on fourth and fifth order Runge-Kutta formulas. As we mentioned in Section 7.2.1, **Method** $\rightarrow$ **ExplicitEuler** leads to the Euler Method. Unlike most methods for **NDSolve**, this choice uses a constant step size, which you can set with the option **StartingStepSize**. The default setting for **NDSolve**, **Method** $\rightarrow$ **Automatic**, specifies the combination of the Adams-Moulton Methods and the Backwards Differentiation Formulas discussed above.

7.4   Round-off Error

The type of error discussed above is called *discretization error*, and would be present even if one could retain an infinite number of digits. In addition, there is *round-off error*, which arises because the computer uses a fixed, finite number of digits. Letting $\tilde{y}_a(t)$ denote the actual computed approximate solution, which includes round-off error, the total error can be written

$$
\begin{aligned}
y(t) - \tilde{y}_a(t) &= (y(t) - y_a(t)) + (y_a(t) - \tilde{y}_a(t)) \\
&= \text{Discretization Error} + \text{Round-off Error}.
\end{aligned}
$$

Since most computers that run *Mathematica* carry 16 digits, the major portion of the error will be the discretization error. Thus, for most problems, round-off error can be safely ignored. For this reason we will not distinguish between $y_a(t)$ and $\tilde{y}_a(t)$.

7.5 Controlling the Error in **NDSolve**

As indicated above, **NDSolve** attempts to provide an approximate solution with error approximately $\leq 10^{-6}$. A more accurate approximate solution can be obtained by using certain options to **NDSolve**. For example,

```
sol4 = NDSolve[{y'[t] == t/y[t], y[0] == 1}, y[t],
    {t, 0, 2}, AccuracyGoal → 15, PrecisionGoal → 15,
    WorkingPrecision → 25]
ya4[t_] = y[t] /. First[sol4]
N[ya4[1.5], 16]
```

produces 1.802775637731983, which is a 14-digit approximation to the exact value $\sqrt{13}/2$ of the solution of problem (7.1) at $t = 1.5$.

AccuracyGoal specifies absolute error, and **PrecisionGoal** specifies relative error. With the settings **AccuracyGoal** → **a** and **PrecisionGoal** → **p**, where **a** and **p** are positive integers, **NDSolve** attempts to calculate $y(t)$ with error less than $10^{-a} + |y(t)|10^{-p}$.

The setting **WorkingPrecision** → **n** fixes the number of digits that **NDSolve** uses in its calculations. Without these options, *Mathematica* uses the default settings:

> **WorkingPrecision** = Normal machine precision for the computer
> **AccuracyGoal** = **PrecisionGoal** = **WorkingPrecision** − 10

So for any computer that has a machine precision of 16 digits, the default options lead to approximately 6-digit accuracy. This will be satisfactory for most of the calculations in this course; but see *Reliability of Numerical Methods* below.

When you use *Mathematica* to calculate an approximation to an initial value problem on an interval $[a, b]$, you should be aware that the solution may not exist over the entire interval. Naturally, *Mathematica* can only calculate the solution over the largest subinterval of $[a, b]$ on which it exists. See the second exercise in this chapter to learn how **NDSolve** reports the answer in this situation.

Now suppose the solution does exist over the entire interval $[a, b]$. In attempting to calculate the solution with the desired accuracy, **NDSolve** will use up to $10,000$ steps.

In certain problems this limit will be reached before the solution has been calculated over the entire interval, and **NDSolve** will return a message indicating this. You can then change the default setting to a higher number, say 20, 000, with the option **MaxSteps** → **20000**, and obtain the solution on a (possibly) larger interval.

Exercise 7.3 Attempt to calculate the solution to

$$\frac{dy}{dt} = y \sin(t^2), \qquad y(0) = 1$$

over the interval $[0, 50]$, first using the default option **MaxSteps** → **10000**, and then using **MaxSteps** → **11000**. What difference do you observe? Find a value for **MaxSteps** that allows the solution to be calculated over the entire interval $[0, 50]$. Graph the solution. (You will probably need to increase **PlotPoints** to get fine enough resolution in the plot.)

7.6   Reliability of Numerical Methods

We have claimed that **NDSolve**, when used with the default options, leads to approximately 6-digit accuracy. More precisely, the default options ensure that the *local error*, *i.e.*, the error made in a step of length h, is approximately 10^{-6}. One is generally more interested in the global error, defined above to be the cumulative error committed in taking the necessary number of steps to get from the initial point t_0 to b. This error cannot be controlled completely by the numerical method because it depends on the differential equation as well as on the numerical method. The key issue is whether the differential equation is stable or unstable (see Chapter 5). In the earlier discussion of local and global error for the Euler Method, we saw that the global error grows only moderately if the differential equation is stable, but grows much faster if the equation is unstable. We will now illustrate the effect of stability and instability on accuracy by considering an example for which **NDSolve** gives poor results.

Example 7.4 Consider the initial value problem

$$\frac{dy}{dt} = y - 3e^{-2t}, \qquad y(0) = 1. \tag{7.7}$$

The exact solution is $y(t) = e^{-2t}$.

Now we find the numerical solution using **NDSolve**, and plot it together with the exact solution. We display the exact solution with a solid line and the approximate solution with a dotted line:

```
sol5 = NDSolve[{y'[t] - y[t] == -3Exp[-2t], y[0] == 1},
    y[t],/,{t, 0, 20}];
ya5[t_] = y[t] /. First[sol5]
Plot[{Exp[-2t], ya5[t]}, {t, 0, 20}, PlotRange → {-1, 1},
```

```
PlotStyle → {Automatic, Dotted}, Axes → False,
Frame → True]
```

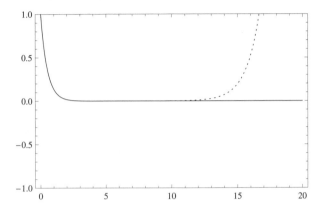

Figure 7.4: Exact (solid line) and Approximate (dotted line) Solutions of Problem (7.7)

We see from Figure 7.4 that the graphs of the approximate and the exact and approximate solutions are indistinguishable from 0 to about 10, but then the curves separate sharply. Their values at $t = 15$ are very different: $0.196\cdots$ *vs.* $e^{-30} = 9.35\cdots \times 10^{-14}$.

How can this failure be explained? The solution to our differential equation with initial condition $y(0) = 1 + \epsilon$ is $y(t) = e^{-2t} + \epsilon e^{t}$. Figure 7.5 shows a plot of the solutions corresponding to initial values $1-(0.5\times 10^{-5}), 1-(0.25\times 10^{-5}), 1, 1+(0.25\times 10^{-5}), 1+ (0.5 \times 10^{-5})$.

We see that as t increases, the solutions with initial values slightly different from 1 separate very sharply from the solution with initial value exactly 1; those with initial values greater than 1 approach $+\infty$, and those with initial values less than 1 approach $-\infty$. Now **NDSolve**, like any numerical method, introduces errors. These errors—whether discretization errors or round-off errors—have caused the numerical solution to jump to a solution that started just above 1. Once on such a solution, the numerical solution will follow it or another such solution. Since these solutions approach $+\infty$, the numerical solution does likewise.

Could we have anticipated this failure? Recall the discussion of stable and unstable differential equations in Chapter 5. The solutions of unstable equations are very sensitive to their initial values, and hence a numerical method will have trouble following the solution it is attempting to calculate. Recall also that an equation $dy/dt = f(t,y)$ is unstable if $\partial f/\partial y > 0$, and stable if $\partial f/\partial y \leq 0$. For our example, $f(t,y) = y - 3e^{-2t}$ and $\partial f/\partial y = 1 > 0$. So we are trying to approximate the solution of an unstable differential equation, and we should not be surprised that we have trouble, especially over long intervals. (By contrast, note that for the initial value problem (7.1) above, $\partial f/\partial y = -t/y^2 < 0$ for

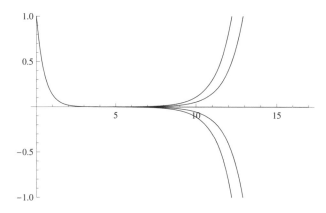

Figure 7.5: Exact Solution with Five Different Initial Values

$t > 0$.) All this suggests caution when dealing with unstable equations. Caution is always recommended when modeling physical problems with unstable equations, especially over long intervals (*cf.* the discussion in Section 5.3).

In assessing the reliability of a numerically computed solution there is another test one can make. Suppose we have made a calculation using the default options, as we have done with our example. Then one can do another calculation, using a more stringent accuracy requirement, and compare the results. If they are nearly the same, one can have reasonable confidence that they are both accurate; but if they differ substantially, then one should suspect that the first calculation is not accurate.

Let's do this with our example. Specifically, we calculate and plot the numerical solution satisfying more stringent accuracy requirements, together with the first numerical solution and the exact solution.

```
sol6 = NDSolve[{y'[t] - y[t] == -3Exp[-2t], y[0] == 1},
    y[x], {x, 0, 20}, AccuracyGoal → 9, PrecisionGoal→ 9
ya6[t_] = y[t] /. First[sol6]
```

We display the exact solution with a solid line, the first numerical solution with a dotted line, and the second with a dashed line:

```
Plot[{Exp[-2t], ya5[t], ya6[t]}, {t, 0, 20},
    PlotRange → {-1, 1},
    PlotStyle → {Automatic, Dotted, Dashed},
    Axes → False, Frame → True]
```

We see from Figure 7.6 that the second numerical solution is accurate out to about 13, whereas the first was accurate out to about 10. In particular, the two numerical solutions differ noticeably after about 10. Even if we didn't know the exact solution, we could have concluded that the first numerical solution is accurate out to, but not beyond, approximately

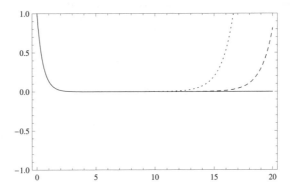

Figure 7.6: Effect of Increased Accuracy

10.

In summary, we have learned that local accuracy—the accuracy that a numerical method can control—leads to reasonable global accuracy if the differential equation is stable, and that the reliability of a numerical solution can also be checked by computing another numerical solution with a more stringent accuracy requirement.

Chapter 8

Features of *Mathematica*

This chapter is a continuation of Chapters 3 and 4. We describe some of the more advanced features of *Mathematica*, focusing on *Mathematica* commands and techniques that are most useful for studying differential equations.

8.1   Functions and Expressions

As we discussed briefly in Section 3.4, *Mathematica* distinguishes between functions and expressions. Strictly speaking, if we define $f(x) = x^2 + 1$, then f (written without any particular input) is a function, whereas $f(x)$ and $x^2 + 1$ are expressions involving the variable x. In mathematical discourse, we often blur this distinction by calling $f(x)$ or $x^2 + 1$ a function, but in *Mathematica* the difference between functions and expressions is important. In order to use *Mathematica* effectively, you must learn to make the same distinction.

The simplest way to define a function in *Mathematica* is via the assignment method. For example, to define $f(x)$ to be the cubic polynomial $x^3 - 2x^2 + 1$, type

```
In[1]:= f[x_] = x^3 - 2 x^2 + 1
```

```
Out[1]= 1 - 2 x^2 + x^3
```

On the left-hand side of the equal sign, the input variable used to define the function must be followed by an underscore and enclosed in square brackets.

A function can be evaluated with either a numeric or a symbolic argument:

```
In[2]:= f[-1]
```

```
Out[2]= -2
```

```
In[3]:= f[u^2]
```

```
Out[3]= 1 - 2 u^4 + u^6
```

The result of this evaluation is the expression (not function) $1 - 2u^4 + u^6$. This is a subtle point: whereas `f` is the name of a function, `f[u^2]` is an expression involving `u^2`.

97

An alternative way of defining a function is to delay evaluation by using the "colon-equal" sign (`:=`). For example,

    In[4] := f[x_] := x^3 - 2 x^2 + 1

The difference between the two methods is the following. In the first (without the colon), the right-hand side of the definition is evaluated at the time the definition is made. In the second (with the colon), the right-hand side of the definition is evaluated only when the value of the function is requested. This is illustrated by *Mathematica*'s lack of response to the second definition, which indicates that *Mathematica* is holding the definition in reserve for evaluation at a later time. Here is an example to illustrate this distinction.

    In[5] := expr = x^2 + 3

    Out[5] = 3 + x^2

    In[6] := g[x_] = expr

    Out[6] = 3 + x^2

    In[7] := h[x_] := expr

    In[8] := g[2]

    Out[8] = 7

    In[9] := h[2]

    Out[9] = 3 + x^2

What happens here is that when *Mathematica* evaluates `h[2]`, it first substitutes `2` for `x` in the expression that defines `h`, and then evaluates the expression. Since `x` does not appear explicitly as a variable on the right-hand side in the definition of `h`, the function we define this way is actually independent of `x`. Thus `h` has been defined to be a function that returns as its *constant* value the expression `expr`. The point of this discussion is that the distinction between `=` and `:=` can be important depending upon when you actually want evaluation to occur. A crucial example appears later when we discuss **NDSolve**.

Finally, we can define functions of several variables. For example, we could define `phi[x_,y_] = x^2 + y^2`.

8.2   Pure Functions

In the examples above, each of the functions we defined was given a name—**f**, **g**, **h**, or **phi**. *Mathematica* provides a way to define pure functions without having to name them. In the previous section, we defined the function f that takes x to $x^3 - 2x^2 + 1$ with `f[x_]` `:= x^3 - 2 x^2 + 1`. Here is a way to define it using a pure function.

    In[10] := #^3 - 2 #^2 + 1 &

Out [10] = #1^3 - 2 #1^2 + 1 &

In the above syntax, the number sign (**#**) represents the first argument supplied to a pure function, and the ampersand (**&**) indicates that the expression is a pure function. The "1" after the number sign in the output is there to indicate the first argument (since there is no second argument in this case, **#1** is equivalent to **#**). Supplying the argument **x** to the pure function yields the function evaluated at x.

In [11] := **% [x]**

Out [11] = 1 - 2 x^2 + x^3

An equivalent way to write a pure function in *Mathematica* is to use the **Function** command.

In [12] := **Function [x, x^3 - 2 x^2 + 1]**

Out [12] = Function [x, x^3 - 2 x^2 + 1]

The command **Function [x, x^3 - 2 x^2 + 1] [-1]** would evaluate the function at -1.

The only time we will use pure functions in this book is as *scale functions* for the **VectorFieldPlot** command; however, *Mathematica* occasionally reports its answers in pure function form. For example, if **f** has been defined as in the preceding section, then *Mathematica* reports the derivative of **f** in pure function format if you type

In [13] := **f'**

Out [13] = -4 #1 + 3 #1^2 &

You can avoid the pure function format in this case by typing

In [14] := **f' [x]**

Out [14] = -4 x + 3 x^2

Finally, we can define pure functions of several variables. For example, the commands **Function [x, y, x^2 + y^2] [a, b]** and **(#1^2 + #2^2) & [a, b]** would both yield the expression a^2 + b^2.

8.3   Clearing Values

One of the most common sources of errors in using *Mathematica* is the failure to keep track of variable or function definitions. For example, suppose that during the course of a *Mathematica* session you type

In [15] := **k = 1;**

You keep working, and twenty minutes later you try to define and evaluate a function of two variables by typing

In [16] := **g [x_, k_] = x + k;**

        g [3, 4]

Out[17]= 4

You might expect the answer to be $3 + 4 = 7$. But in the definition of **g**, Mathematica evaluates **x + k** and substitutes the value 1 for **k**, so effectively **g[x, k]** is defined to be **x + 1**. In a complicated expression, this type of unintended substitution can be difficult to detect.

The simplest way to avoid this problem is to clear a variable either before using it or immediately after using it. For example, you can ensure that **g** is defined as intended in the preceding example by typing

    In[18] := **Clear[g, x, k]**

             g[x_, k_] = x + k

Out[19]= k + x

Another way to avoid unintended substitution is by using **:=** instead of **=**.

    In[20] := **k = 1;**

             g[x_, k_] := x + k

             g[3, 4]

Out[22]= 7

In this example, because we have used **:=** to define the function, *Mathematica* does not evaluate the right-hand side until we type **g[3, 4]**, at which point it substitutes the value 4 for **k**. For this reason, it is often safer to use **:=** rather than **=** when defining functions.

If you typed something like **q'[t] = t^2** (which is easy to do accidentally when using **DSolve**), then **Clear[q']** doesn't work, but **Clear[q]** does.

You can always find out the current value of a function or variable by typing a question mark (**?**) before the function or variable name. For example, **?f** will show you the current value or definition of **f**.

You should be aware that if you have several Notebooks open in a single *Mathematica* session, the definitions you make in one Notebook will carry over to the other Notebooks. This happens because the underlying *Mathematica* process, or *kernel*, is the same for all the Notebooks. This also accounts for the fact that In and Out numbering carries over from Notebook to Notebook in a single *Mathematica* session; if you enter input in one Notebook and the input is labeled In[n], then the next input will be labeled In[n + 1], even if it is in a different Notebook.

8.4   Managing Memory

We have noted that *Mathematica* stores in computer memory *every* input statement and *every* output (that's what it's doing when it assigns those labels In[n] and Out[n]). You can check the number of bytes currently being used to store all data in the current *Mathematica* kernel session by typing and evaluating **MemoryInUse[]**.

In a long *Mathematica* session, you can use a lot of computer memory and this might lead to a crash if the available memory is exhausted. The simplest way to ameliorate this problem is to reset the line numbering to zero. To do this, simply type **$Line = 0**. The In/Out numbering will be reset, and subsequent inputs and outputs will be written over the old ones instead of using new memory space.

Another way to free up memory is to quit and restart the kernel. We described how to do this in Chapter 4. You should probably quit and restart the kernel each time you start a new problem.

8.5 The Replacement Operator and Transformation Rules

We have seen that a function can be evaluated at various values of the independent variable. A similar effect can be accomplished with expressions by using the *replacement operator* **/.** (pronounced "slash-dot"), together with a *transformation rule*. For example,

In[23] := **x^2 - 1 /. x → 5**

Out[23] = 24

This input means: in the expression $x^2 - 1$, replace x by 5. The notation **x → 5** is an example of a transformation rule.

The replacement operator is extremely useful and versatile. What it does is to substitute for one or more variables in an expression using one or more transformation rules. Here is another example:

In[24] := **(x^2 - 3 y) / z /. {x → a, y → b, z → c}**

Out[24] = $\dfrac{a^2 - 3 b}{c}$

A general feature of *Mathematica*'s various "solve" routines is that the solutions are reported in the form of transformation rules. For example, the simple quadratic equation $x^2 - 1 = 0$ has the two solutions $x = 1$ and $x = -1$.

In[25] := **sol = Solve[x^2 - 1 == 0, x]**

Out[25] = **{{x → -1}, {x → 1}}**

Mathematica's response is a list consisting of two transformation rules. (We have named this list **sol** so we can use it later.) Note that *Mathematica* does not report these solutions in the form "$x = 1$" and "$x = -1$", because those two assignments would be in conflict. In fact, *Mathematica* *never* assigns values to variables unless specifically directed to do so. Transformation rules are *Mathematica*'s way of steering clear of assigning values to variables.

Now let us use the transformation rules that we generated with the **Solve** command. To do this, we have to extract the transformation rules from the list. If we want to use the rule **x → -1** on the expression $x^2 + x$, we type

In[26] := **x^2 + x /. First[sol]**

Out[26] = 0

To use the rule $\mathbf{x} \rightarrow \mathbf{1}$, we type

```
In[27]:= x^2 + x /. Last[sol]
```

```
Out[27]= 2
```

The most common use of the replacement operator is to transform expressions that have been generated previously by *Mathematica* (see Section 5.1).

8.6 Equations *vs.* Assignments

In *Mathematica*, an *equation* is different from an *assignment*. For example, $\mathbf{x\ ==\ 2}$ is an equation, whereas $\mathbf{x\ =\ 2}$ or $\mathbf{x\ :=\ 2}$ is an assignment. Compare the effects of entering an equation and an assignment.

```
In[28]:= x == 2
```

```
Out[28]= x == 2
```

```
In[29]:= x = 2
```

```
Out[29]= 2
```

Note that an equation has no effect on the value of the variables in the equation. Names can even be assigned to equations.

```
In[30]:= niceeqn =   energy == m*c^2
```

```
Out[30]= energy == c² m
```

8.7 Differentiation

The distinction between functions and expressions becomes important when differentiating. *Mathematica* has two commands for differentiating. The **D** command operates on expressions.

```
In[31]:= Clear[g, x]
          g[x_] := x^2
          D[g[x], x]
```

```
Out[33]= 2 x
```

Notice that the output of **D** is another expression. The syntax for second derivatives is $\mathbf{D[g[x], \{x, 2\}]}$, and for nth derivatives, $\mathbf{D[g[x], \{x, n\}]}$. When differentiating an expression with **D**, you must state explicitly which variable to use.

As we observed in the section on pure functions, the *prime* operator (**'**) differentiates a function and returns a pure function. One place where it is best to use the prime format is to specify initial values for derivatives in **DSolve** or **NDSolve**. So, to solve the second order initial value problem $y'' = 9y + 2x$, $y(0) = 1$, $y'(0) = 0$, you could type

```
In[34]:= diffeqn = D[y[x], {x, 2}] == 9 y[x] + 2 x
```

```
Out[34]= y''[x] == 2 x + 9 y[x]
```

```
In[35]:= initialpos = y[0] == 1;
         initialvel = y'[0] == 0;
```

```
In[37]:= DSolve[{diffeqn, initialpos, initialvel}, y[x], x]
```

$$\text{Out[37]= } \left\{\left\{y[x] \rightarrow \frac{1}{54} e^{-3x} (25 + 29 e^{6x} - 12 e^{3x} x)\right\}\right\}$$

8.8 Vectors and Matrices

A *vector* is an ordered list of numbers. If the numbers are listed horizontally, the vector is called a row vector; if they are listed vertically it is a column vector. Vectors in *Mathematica* are simply represented by lists. For example:

```
In[38]:= v1 = {1, 2, 3}
```

```
Out[38]= {1, 2, 3}
```

```
In[39]:= v2 = {4, 6, 5, 2, 8}
```

```
Out[39]= {4, 6, 5, 2, 8}
```

Because of the way *Mathematica* uses lists to represent vectors, you never have to distinguish between row and column vectors.

A *matrix* is a rectangular array of numbers. Row and column vectors can be viewed as special classes of matrices. Matrices in *Mathematica* are represented by lists of lists. Consider the 3×4 matrix

$$m = \begin{pmatrix} 1 & 2 & 3 & 4 \\ 5 & 6 & 7 & 8 \\ 9 & 10 & 11 & 12 \end{pmatrix}.$$

It can be entered in *Mathematica* with the command

```
In[40]:= m = {{1, 2, 3, 4}, {5, 6, 7, 8}, {9, 10, 11, 12}}
```

```
Out[40]= {{1, 2, 3, 4}, {5, 6, 7, 8}, {9, 10, 11, 12}}
```

We can display **m** in standard matrix form using the **MatrixForm** command.

```
In[41]:= MatrixForm[m]
```

```
Out[41]//MatrixForm=
```

$$\begin{pmatrix} 1 & 2 & 3 & 4 \\ 5 & 6 & 7 & 8 \\ 9 & 10 & 11 & 12 \end{pmatrix}$$

The dimensions of a matrix can be found with **Dimensions**:

In[42]:= **Dimensions[m]**

Out[42]= {3, 4}

If two matrices **A** and **B** are the same size, their sum is obtained by typing **A + B**. You can also add a scalar (a single number) to a matrix; **A + c** adds **c** to each element in **A**. Likewise, **A - B** represents the difference of **A** and **B**, and **A - c** subtracts the number **c** from each element of **A**. If **A** and **B** are multiplicatively compatible, *i.e.*, if **A** is $n \times m$ and **B** is $m \times \ell$, then their product **A . B** is $n \times \ell$. The "dot operator" (**.**) also gives the scalar product of two vectors, and multiplies a matrix by a vector. Since *Mathematica* makes no distinction between row and column vectors, you can use "dot" for both left- and right-multiplication of vectors by matrices. The product of a number **c** and the matrix **A** is given by **c * A**, and **Transpose[A]** represents the transpose of **A**.

If **A** and **B** are the same size, then **A * B** is the *element-by-element* product of **A** and **B**, *i.e.*, the matrix whose i, j element is the product of the i, j elements of **A** and **B**. Likewise, **A / B** is the element-by-element quotient of **A** and **B**, and **A^c** is the matrix formed by raising each of the elements of **A** to the power **c**. More generally, if **f** is one of the built-in mathematical functions in *Mathematica*, then **f[A]** is the matrix obtained by applying **f** element-by-element to **A**. See what happens when you type **Sqrt[m]**, where **m** is the matrix defined above.

Recall that **x[[3]]** is the third element of a vector **x**. Likewise, **A[[2, 3]]** represents the $2, 3$ element of **A**, *i.e.*, the element in the second row and third column. You can specify submatrices in a similar way. Typing **A[[2, {2, 4}]]** yields the second and fourth elements of the second row of **A**. To select the second, third, and fourth elements of this row, type **A[[2, 2;;4]]**. The submatrix consisting of the elements in rows 2 and 3 and in columns 2, 3, and 4 is generated by **A[[2;;3, 2;;4]]**. You can also select an entire row or column. For example, **A[[2]]** picks out the second row of **A**, and **A[[All, 3]]** yields the third column of **A**.

Mathematica has several commands that generate special matrices. The command **Array[f, {n, m}]** generates an $n \times m$ matrix whose i, j element is $f[i, j]$. For example, to produce an $n \times m$ matrix of zeros or ones you can type **Array[0 &, {n, m}]** or **Array[1 &, {n, m}]**, respectively. (Note the use of constant pure functions here.) Also, **IdentityMatrix[n]** represents the $n \times n$ identity matrix.

8.8.1   Solving Linear Systems

Suppose **A** is a nonsingular $n \times n$ matrix and **b** is a vector of length n. Then typing **LinearSolve[A, b]** numerically computes the unique solution to **A.x = b**. Type **?LinearSolve** for more information.

If either **A** or **b** is symbolic rather than numeric, then **LinearSolve[A, b]** computes the solution to **A.x = b** symbolically. **LinearSolve** also works when the right-hand side of the matrix equation is also a matrix. If no right-hand side vector is specified, **LinearSolve** will generate a **LinearSolveFunction**, which can be applied repeatedly to different **b**'s. Here is an example to illustrate the use of **LinearSolve** and **LinearSolveFunction**.

```
In[43]:= Clear[b, f, m]
          m = {{1, 2},{3, 4}};
          b = {5, 6};
          LinearSolve[m, b]
```

$$Out[46] = \left\{ -4, \frac{9}{2} \right\}$$

Alternatively, we can type:

```
In[47]:= f = LinearSolve[m]
```

```
Out[47]= LinearSolveFunction[{2, 2}, <>]
```

```
In[48]:= f[{5, 6}]
```

$$Out[48] = \left\{ -4, \frac{9}{2} \right\}$$

8.8.2 Calculating Eigenvalues and Eigenvectors

Mathematica has various functions for computing eigenvalues and eigenvectors of a square matrix **A**. The command **Eigenvalues[A]** will generate a list of the eigenvalues of **A**, **Eigenvectors[A]** will generate a list of the eigenvectors of **A**, and **Eigensystem[A]** will produce the same output as {**Eigenvalues[A]**, **Eigenvectors[A]**}. Here are three examples illustrating the use of **Eigenvalues**, **Eigenvectors**, and **Eigensystem**.

```
In[49]:= m = {{3, -2, 0}, {2, -2, 0}, {0, 1, 1}}
```

```
Out[49]= {{3, -2, 0}, {2, -2, 0}, {0, 1, 1}}
```

This matrix has three eigenvalues.

```
In[50]:= Eigenvalues[m]
```

```
Out[50]= {2, -1, 1}
```

Here are the three eigenvectors of **m**.

```
In[51]:= Eigenvectors[m]
```

```
Out[51]= {{2, 1, 1}, {-1, -2, 1}, {0, 0, 1}}
```

Here are the eigenpairs of **m**. We will assign the list of eigenvalues to **vals** and the list of eigenvectors to **vecs**.

```
In[52]:= {vals, vecs} = Eigensystem[m]
```

```
Out[52]= {{2, -1, 1}, {{2, 1, 1}, {-1, -2, 1}, {0, 0, 1}}}
```

The first eigenvalue given in the first list (**vals**) corresponds to the first eigenvector in the second list (**vecs**), and so on.

8.9   Graphics

In Section 3.7 we discussed the plotting commands **Plot**, **ListPlot**, **ContourPlot**, **ParametricPlot**, and **Show**, as well as the basic commands for adjusting and labeling the axes of a graph. These commands, along with **VectorFieldPlot**, which we introduced in Chapter 6 to plot direction fields, are the main graphics commands you will need for this course. In this section we describe additional commands for manipulating graphics.

The most important thing to keep in mind for the various plotting commands is the type of *Mathematica* objects that the command is designed to plot. For example, **Plot** plots expressions involving a single variable. On the other hand, **VectorFieldPlot** plots a list of two expressions involving two variables. If *Mathematica* gives you an error message or an implausible plot when you enter a plotting command, the first things to check are whether you are plotting the right kind of object for that particular command, and whether you are using precisely the right syntax for that object and that command.

8.9.1   Options for Plots

Plot is the simplest and most versatile plotting command; it plots an expression or list of expressions involving a single variable over a specified interval. The vertical range of the plot can also be specified using the option **PlotRange**. For example, to plot the two functions $e^x \cos x$ and $x^3 - 3x + 2$ over the interval $[-3, 3]$ with range $[-10, 10]$, type:

```
In[53]:= plot1 = Plot[{Exp[x] Cos[x], x^3 - 3x + 2},
              {x, -3, 3}, PlotRange → {-10, 10}]
```

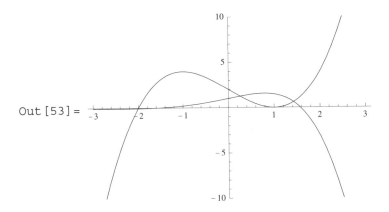

Out[53] =

Note that we have named this plot "**plot1**" for future reference.

When *Mathematica* responds to a **Plot** command, it has to make many choices: what scales to use, where the function should be sampled, how the axes should be drawn, *etc.* Most of the time, *Mathematica* makes good choices. However, there may be times when you want to adjust *Mathematica*'s default settings. You can do this by modifying the values of one or more of **Plot**'s many options. Usually you will only need to use a few of them at a time. Each time you produce a plot, you can specify the options you wish to modify by including a rule or sequence of rules in the form *optionname* → *value* as the last argument(s) to the **Plot** function. The most common options you will encounter are explained below. See the *Mathematica* Documentation Center for more details on these and other options.

In Section 3.7.2, we introduced **PlotRange, AspectRatio, PlotLabel**, and **AxesLabel**. We will now say a bit more about these, and also introduce other options. **PlotRange** can be used to adjust both the horizontal and vertical range of the plot. The syntax is **PlotRange** → $\{\{x_{min}, x_{max}\}, \{y_{min}, y_{max}\}\}$. This will restrict the plot to the rectangle $x_{min} \leq x \leq x_{max}, y_{min} \leq y \leq y_{max}$.

To specify whether axes should be drawn, use **Axes**. The default is **Axes** → **True**, which draws all axes. **Axes** → **False** draws no axes and **Axes** → **{False, True}** draws a vertical axis but no horizontal axis.

To specify whether to draw a frame around a two-dimensional plot use **Frame**. The default setting is **Frame** → **False**. **Frame** is analogous to **Axes** in that **Frame** → **True** draws a frame around the whole plot and **Frame** → **{False, True}** draws a frame along the left and right edges but none along the top and bottom.

AxesOrigin is an option for two-dimensional plots that specifies the point at which axes should cross. The default setting, **AxesOrigin** → **Automatic**, uses an internal algorithm to determine where the axes should cross. If the point $(0,0)$ is within, or close to, the plotting region, then it is usually chosen as the axes' origin. However, some *Mathematica* plots can be misleading if you do not pay attention to where the axes cross. For example, the input **Plot[x^3 - x + 1, {x, -1, 1}]** will produce a graph with axes that cross at the point $(0, 0.6)$. To produce a plot with axes that cross at $(0,0)$ type:

```
In[54] := plot2 = Plot[x^3 - x + 1, {x, -1, 1},
              AxesOrigin → {0, 0}]
```

Note that we have named this plot, which appears on the next page, "**plot2**" for future reference.

To specify labels for axes, frames, or an overall label for a plot use **AxesLabel**, **FrameLabel** and **PlotLabel**, respectively. **AxesLabel** → *label* specifies a label for the vertical axis in two dimensions and **AxesLabel** → $\{label_x, label_y\}$ specifies labels for both the horizontal and vertical axes. The use of **FrameLabel** is analogous to **AxesLabel**. **PlotLabel** → *label* specifies a label for the plot. The expressions you give as labels are printed just as they would be if they appeared as *Mathematica* output. You specify arbitrary text by putting it inside a pair of double quotes, like this: **"text"**.

To specify tick marks for axes or frames use **Ticks** and **FrameTicks**, respectively. The default setting for both of these options is **Automatic** when either axes or frames

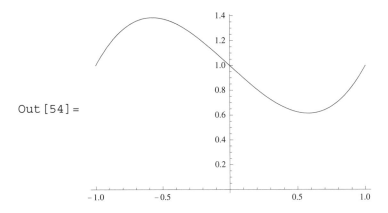

Out [54] =

are drawn. By default, *Mathematica* usually places tick marks at points whose coordinates have the minimum number of digits in their decimal representation.

To get smooth curves, *Mathematica* has to evaluate functions you plot at a large number of points. In general, *Mathematica* will *adapt* its sampling of points to the form of the function. As a result, in places where a function is exceptionally oscillatory, *Mathematica* will use more points. However, *Mathematica* can only sample functions at a limited number of points, and therefore it may miss something. You can force *Mathematica* to sample a function at a larger number of points with the option **PlotPoints**. The default setting for **PlotPoints** depends on the plotting command; for **Plot** the default is 50. Higher settings may improve your graphs, at the cost of an increase in computation time.

8.9.2   Plotting Vector Fields

VectorFieldPlot is used for plotting vector or direction fields and was described in Chapter 6. It is in the **VectorFieldPlots** package. The **ScaleFunction** option to the **VectorFieldPlot** command can be used to rescale the length of vectors in the plot. The default setting causes the vector field to be displayed with vectors having length proportional to the magnitude of the actual vectors. But often one is interested in the behavior of a vector field near critical points, where the vectors are very short. The default setting produces a plot that is relatively uninformative near critical points, because the vectors are so short that it is impossible to discern their direction. The simplest solution is to make all the non-zero vectors the same size. The option **ScaleFunction → (1 &)** does this. The ampersand is present because **ScaleFunction** expects a *pure function* as its argument; in this case we have used the constant function 1. Note that the default setting of **AspectRatio** is **Automatic** for **VectorFieldPlot**; this can result in a thin graph if the x and y ranges of the plot have different lengths. In such cases you can use the option **AspectRatio → 1** to ensure a square graph.

8.9.3   Redrawing Plots

In Section 3.7.3 we showed how to combine plots by assigning names to the plots and then displaying them together using the **Show** function. You can also use **Show** to redraw plots with some of their options modified. By using **Show** with a sequence of different options, you can look at the same plot in many different ways. All of the options listed above in Section 8.9.1 can also be used for **Show**. For example, to redraw **plot2** above with the specified vertical range $[-2, 2]$ and to make the plot square, type:

In[55]:= **Show[plot2, PlotRange → {-2, 2}, AspectRatio → 1]**

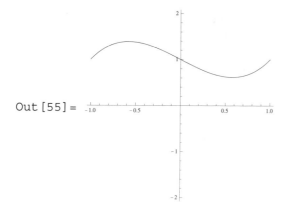

Out[55]=

8.9.4   Editing Figures

In Sections 3.7.2 and 8.9.1 we showed how to use options to modify graphs. Here we describe several other ways to modify and annotate graphs.

When plotting more than one curve, it is often important to distinguish among them by making them have different colors or styles (*e.g.*, solid versus dashed). When plotting multiple curves with **Plot**, *Mathematica* will choose a different color for each curve automatically. Once you've plotted a curve, you can change its appearance in the Notebook as follows. First, double-click the background of the graph to select it for editing; the frame highlight around the graph will be dashed. Then, click on a curve; the curve will be highlighted in orange and a frame highlight will appear around the curve. (If something else gets highlighted, try again.) Next, open the Graphics Inspector palette by selecting **Graphics ▶ Graphics Inspector** or by typing its keyboard shortcut CTRL+G. You can change the color, width, or style of this curve, among other things, with this palette. However, these modifications are only temporary. If you reevaluate the input cell, your changes will be overwritten by the new output cell. See the "Editing *Mathematica* Graphics" tutorial in the online Documentation Center for other options.

An alternative approach is to explicitly choose the line style or color in advance. You can explicitly style your curves by adding the **PlotStyle** option to your plotting command. For example, to produce the plot of $e^x \cos x$ and $x^3 - 3x + 2$ from Out[53] with

$e^x \cos x$ as a dashed, blue curve and $x^3 - 3x + 2$ as a solid, red curve, add the argument **PlotStyle** → **{{Dashed, Blue}, Red}**. **PlotStyle** can apply to points, lines and surfaces. You can also add a legend to your plot to identify curves with the option **PlotLegend**, which can be found in the **PlotLegends** package. These options cannot be used in **Show**. See the "Graphics and Sound" tutorial in the online Documentation Center for more information.

Another effective approach when presenting multiple curves on the same graph is to label the curves directly. You can add text, arrows, and other annotations using the Drawing Tools palette. To open the Drawing Tools palette type CTRL+D or select **Graphics ▶ Drawing Tools**. Now select the type of annotation you want, then click the spot on the graph where you want it to appear. For text (plain text or traditional mathematical notation), type the text, while for a line or arrow, drag the pointer to draw the arrow. Alternatively, to add text or a symbol exactly at specific coordinates—say, the coordinates of a special point you want to highlight on a curve—you can use the **Text** and **Graphics** commands. For example, to annotate **plot1** from above with the expressions $x^3 - 3x + 2$ and $e^x \cos(x)$ at the coordinates $(-1, 5)$ and $(1, 3)$, respectively, type:

```
In[56]:= t1 = Graphics[Text[x^3 - 3x + 2, {-1, 5}]];

        t2 = Graphics[Text[Exp[x] Cos[x], {1, 3}]];

        Show[plot1, t1, t2]
```

Out[57] =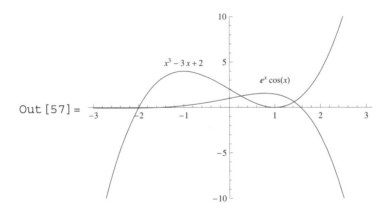

Finally, by selecting a graph with the left mouse button, you can move the graph, resize it, *etc.* First, click anywhere on the background of the graph to select it. You can then resize the graph by dragging one of the small squares (called handles) along the edges that appear after you have selected the graph. Hold down the SHIFT key while dragging a handle to change the aspect ratio, or hold down the CONTROL key to crop the plot. You can move the graph by dragging the frame (not a handle). This will also push out the margins around the graph. If you wish to return your Graphics cell to its default size and alignment, select the cell (by clicking on its bracket), and choose **Graphics ▶ Rendering ▶ Make Standard Size**.

8.10 The **Evaluate** Command

As mentioned in Section 8.9.1, when you ask *Mathematica* to plot an expression $f(x)$, *Mathematica* has to evaluate the function f at a large number of points in order to produce a smooth curve. There are two possible approaches *Mathematica* can use to do this. One approach is first to try to evaluate $f(x)$ symbolically to get a simpler expression in terms of x, and then to evaluate the simpler expression numerically for the specific values of x needed to produce the plot.

The second approach is to evaluate $f(x)$ from scratch for each value of x chosen by the plotting routine. This is the approach that *Mathematica* plotting commands use by default. However, there are times when it is better to use the first approach, which you can do by applying the **Evaluate** command to the expression being plotted. Using **Evaluate** in this way is generally more efficient computationally, and it can sometimes avoid errors or unexpected results. An example is when f is actually a command that generates a list of functions, such as **Table**, which we used in Chapters 5 and 7 to plot a family of solutions to a differential equation. If you type **Plot[Evaluate[Table[x^c, {c, 1, 5}]], {x, -5, 5}]**, *Mathematica* will produce a plot of five different curves, each with its own distinctive color. This is the default coloring scheme for plotting a list of expressions. However, if you type **Plot[Table[x^c, {c, 1, 5}], {x, -5, 5}]**, *Mathematica* will produce the same five curves but all in the same color.

Here is another example that demonstrates the use of the **Evaluate** command.

```
In[58]:= Plot[Evaluate[Table[BesselJ[n, x], {n, 0, 3}]],
            {x, 0, 10},
            PlotStyle → {Thick, Dashed, Dotted, DotDashed}]
```

Out[58]=

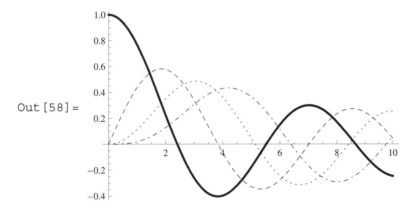

See what happens if you omit **Evaluate** from the above plotting routine. Note that whereas in this case one still gets a plot, there will be cases later in this book where omitting **Evaluate** from a plotting command using **Table** causes it to take a huge amount of time, or even to fail to produce any output at all.

8.11 More about `DSolve`

We introduced `DSolve`, *Mathematica*'s symbolic ODE solver, in Chapter 5. Here is another example illustrating the use of `DSolve`, in this case to find the general solution of a second order differential equation.

```
In[59] := gensol = DSolve[y''[x] - y[x] == 0, y[x], x]
```

$$\text{Out}[59] = \{\{y[x] \rightarrow e^x\, C[1] + e^{-x}\, C[2]\}\}$$

You should recognize from the discussion earlier in this chapter that *Mathematica* has reported the solution as a *nested list* of transformation rules, even though there is only one element in the list. The output is a general solution, and the expressions `C[1]` and `C[2]` are arbitrary constants. These constants are "protected" variables, so we can't actually assign values to them. But we can replace them with particular values by using the replacement operator.

```
In[60] := ygen[x_] = y[x] /. First[gensol];
          y0[x_] = ygen[x] /. {C[1] → 1, C[2] → -1}
```

$$\text{Out}[61] = -e^{-x} - e^x$$

8.12 More about `NDSolve`

The command `DSolve` yields *symbolic* solutions of differential equations, *i.e.*, algebraic expressions that satisfy the original differential equation. On the other hand, `NDSolve`, which we introduced in Chapter 7, generates approximate *numerical* solutions of differential equations. These solutions are really just lists of numbers corresponding to approximate values of the solution at particular values of the independent variable. However, the output of `NDSolve` can generally be treated as a function, because *Mathematica* interpolates between values in the list. The essential difference you have to keep in mind when using `NDSolve` (as opposed to `DSolve`) is that `NDSolve` requires numerical initial conditions, whereas `DSolve` can work with initial conditions that are variables. The same is true of parameters in the differential equation; for `DSolve` the parameters can be variables, but for `NDSolve` you must supply specific numbers.

```
In[62] := approxsol = NDSolve[{y'[x] == Sqrt[y[x] + x],
             y[1] == 3}, y[x], {x, 1, 10}]
```

$$\text{Out}[62] = \{\{y[x] \rightarrow \text{InterpolatingFunction}[\{\{1., 10.\}\}, <>][x]\}\}$$

The output is different from that of `DSolve`—the transformation rule specifies an interpolating function rather than a specific algebraic expression. But we can manipulate the output in exactly the same way as we did for `DSolve`.

```
In[63] := ya[x_] = y[x] /. First[approxsol]
```

```
Out[63]= InterpolatingFunction[{{1., 10.}}, <>] [x]
```

This defines a function **ya**, which is an approximation to the actual solution of the initial value problem. Note that this function is defined on the interval $[1, 10]$, which is the interval we specified when we executed the **NDSolve** command. We can plot it (with a command like **Plot[ya[x], {x, 1, 10}]**), or evaluate it at specific points (*e.g.*, **ya[3.45]** or **Table[ya[1 + i], {i, 0, 9}]**). *Mathematica* will extrapolate the function if you try to plot it or evaluate it outside the domain of definition.

It is possible to have the solution function depend on the initial data, but here the difference between **NDSolve** and **DSolve** becomes important. Since **NDSolve** works numerically rather than symbolically, it must have specific initial data to do its computations. One approach is as follows:

```
In[64]:= f[c_] :=
         y[x] /. First[NDSolve[{y'[x] == Sqrt[y[x] + x],
         y[1] == c}, y[x], {x, 1, 10}]]
```

Note that **f** is defined with **:=**. This is necessary to prevent *Mathematica* from evaluating **NDSolve** before specific initial data are supplied. When **f** is evaluated for a value of **c**, it will return an expression involving **x**, which could be plotted or evaluated using **/.**.

Now, to plot several solution curves corresponding to different initial data, you can type:

```
In[65]:= Plot[Evaluate[Table[f[c], {c, 1, 5}]],
         {x, 1, 10}]
```

Out[65]=

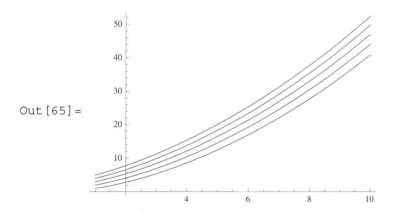

To compute values of **f** for specific choices of the initial data, *e.g.*, to evaluate the value of the solution function at $x = 2$ for initial data $c = 5$, you simply type **f[5] /. x → 2**.

8.12.1   Event Detection

Sometimes the main question of interest in a problem that is modeled by a differential equation is when a specific event occurs. Mathematically, an event is formulated as an algebraic equation that is true when the event occurs. In **NDSolve**, the **EventLocator** method is a built-in controller method which detects events and takes action when they are detected. A simple example of event detection is determining the time required for a pendulum to complete one cycle in its motion and stopping the integration at that point. Consider the simple pendulum equation described in Problem 3 of Problem Set D,

$$\theta'' + \sin\theta = 0.$$

The following command will integrate the pendulum equation numerically with initial conditions $\theta(0) = 1$, $\theta'(0) = 0$ up to the next point at which the event $\theta'(t) = 0$ occurs.

```
In[66]:= psol1 = NDSolve[{θ''[t] + Sin[θ[t]] == 0,
            θ[0] == 1, θ'[0] == 0}, θ, {t, 0, ∞},
            Method → {EventLocator, "Event" → θ'[t]}]
```

```
Out[66]= {{θ → InterpolatingFunction[{{0., 3.34999}}, <>]}}
```

When using the **EventLocator** method, you must set the **"Event"** option to be an expression (involving the solution of the differential equation) that is zero when the event occurs. The default action to take when an event is detected is to stop the integration. From the output we see that $\theta'(t) = 0$ at about $t = 3.35$.

We can extract the final time from the **InterpolatingFunction** object by using the **InterpolatingFunctionAnatomy** package. The following commands determine the time at which the event occurs and plot the solution in the phase plane up to that time.

```
In[67]:= <<DifferentialEquations`InterpolatingFunctionAnatomy`
```

```
In[68]:= domain = InterpolatingFunctionDomain[
            First[θ /. psol1]]
         end = Last[First[domain]];
         ParametricPlot[{θ[t], θ'[t]} /. psol1, {t, 0, end}]
```

```
Out[68]= {{0., 3.34999}}
```

From the graph, which appears on the next page, we see that the integration has only gone over half the period of the pendulum, as it swings from $x = 1$ to $x = -1$. Events are detected when the sign of the event expression changes. We can make the event occur after a full period by restricting the event to be only for a sign change from positive to negative. We do this with the **"Direction"** option.

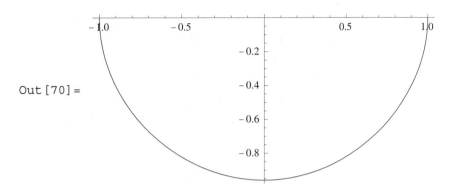

Out[70] =

In[71] := **psol2 =**
  NDSolve[{θ''[t] + Sin[θ[t]] == 0, θ[0] == 1,
    θ'[0] == 0}, θ, {t, 0, ∞},
  Method $\rightarrow$ {EventLocator, "Event" $\rightarrow$ θ'[t],
  "Direction" $\rightarrow$ -1}]

Out[71] = {{$\theta \rightarrow$ InterpolatingFunction[{{0., 6.69998}}, <>]}}

Now we can plot the solution in the phase plane over the full period:

In[72] := **newend =**
  Last[First[InterpolatingFunctionDomain[
    First[θ /. psol2]]]];
  ParametricPlot[{θ[t], θ'[t]} /. psol2,
  {t, 0, newend}]

Out[73] =

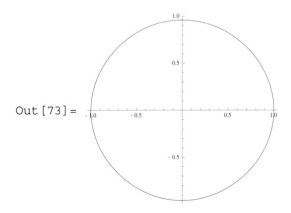

Finally, we can print the time and values each time the event $\theta'(t) = 0$ is detected for $0 \le t \le 50$ in the direction indicated:

```
In[74]:= NDSolve[{θ''[t] + Sin[θ[t]] == 0, θ[0] == 1,
           θ'[0] == 0}, θ, {t, 0, 50},
         Method → {EventLocator, "Event" → θ'[t],
           "Direction" → -1,
           "EventAction" :> Print["θ'[", t, "] = ",
             θ'[t]]}]
```

$\theta'[6.69998] = -2.19963 \times 10^{-15}$

$\theta'[13.4] = -4.1564 \times 10^{-15}$

$\theta'[20.0999] = -8.32667 \times 10^{-15}$

$\theta'[26.7999] = -7.68136 \times 10^{-15}$

$\theta'[33.4999] = -1.06581 \times 10^{-14}$

$\theta'[40.1999] = 1.70003 \times 10^{-14}$

$\theta'[46.8998] = -1.64729 \times 10^{-14}$

```
Out[74]= {{θ → InterpolatingFunction[{{0., 50.}}, <>]}}
```

Note that in this example, the **EventAction** option was given using **RuleDelayed** (**:>**) to prevent the **Print** command from being evaluated until the event is located.

Remark 8.1 Events are detected when the derivative is only approximately zero. The event locator method detects the presence of an event by detecting a sign change of the event expression during a step of the underlying numerical integrating technique. It then uses a numerical method to approximate the time of the event within that step. Since the location process is numerical, you should expect only approximate results. The method will not work if the **"Event"** expression vanishes without changing sign. For example, **NDSolve[..., Method → {EventLocator, "Event" → Cos[t] - 1}]** will not work, since the function $\cos(t) - 1$ never changes sign.

8.13   Troubleshooting

In this section, we list some advice for avoiding common mistakes. We also discuss warning messages and describe some techniques for recovering from errors.

8.13.1   The Most Common Mistakes

Here are some of the most common mistakes in using *Mathematica*.

1. Failing to clear values. See Section 8.3, *Clearing Values*, above.

2. Failing to keep track of In/Out numbering. This mistake sometimes occurs if you use **%** to refer to the output of the previous command. The previous command is the one

most recently executed, not necessarily the preceding command in the Notebook. See Section 3.3.3, *Referring to Previous Output*, in Chapter 3.

3. Using a command from a *Mathematica* package before the package has been loaded. See Section 3.9, *Packages*, in Chapter 3.

4. Typing the wrong kind of "=" sign. The worst instance is typing "=" when you intend "==". This causes a value to be assigned to a variable, and this value often must be explicitly cleared before you can proceed. Be particularly wary of this mistake when using **Solve**, **DSolve**, and **NDSolve**, all of which require an equation as one of the arguments.

5. Mismatching braces or brackets. *Mathematica* usually catches this kind of mistake.

6. Improper use of built-in functions. For example, to use the natural logarithm function in *Mathematica*, you type **Log[x]** and not **Log x** or **Log(x)**.

7. Improperly concatenating variables. For example, **xy** is a variable named xy, whereas **x y** or **x*y** means x *times* y.

8. Using **NDSolve** without numerical initial conditions or without giving a range of values for the independent variable.

9. Omitting the **Evaluate** command when using **Table** inside a plotting command. See Section 8.10, *The* **Evaluate** *Command*, in this chapter.

10. Using lower-case instead of upper-case letters in *Mathematica* commands. For example, *Mathematica* does not recognize **Findroot** or **findroot**; the correct command is **FindRoot**.

11. Trying to plot an expression that cannot be evaluated numerically. In this case, *Mathematica* will produce an empty graph.

8.13.2 Error and Warning Messages

You should pay close attention to the error and warning messages that *Mathematica* generates. Although these messages may seem cryptic at first, you will find that they give valuable clues to locating mistakes in input. For example, suppose that we execute the following command.

```
In[75]:= NDSolve[{y'[x] == Sqrt[y[x] + x], y[1] == 3},
            y[x], x]
```

NDSolve::ndlim : Range specification x is

not of the form {x, xend} or {x, xmin, xmax}. ≫

$$Out[75]= NDSolve[\{y'[x] == \sqrt{x + y[x]}, y[1] == 3\}, y[x], x]$$

This error message points to the source of the problem. The third argument in **NDSolve** must be a list that gives the name of the independent variable and the range of that variable. This is a common error because **NDSolve** and **DSolve** have similar, but not identical, syntaxes.

It is important to distinguish between *error* messages and *warning* messages. An error message means that something is really wrong, and *Mathematica* will probably not be able to produce output. A warning message is an indication that something might be wrong or that *Mathematica* can't do exactly what you've instructed it to do. Here are two examples:

```
In[76]:= sol[x_] =
             y[x] /.
               First[NDSolve[{y'[x] == y[x]^2, y[0] == 1},
                  y[x], {x, 0, 2}]]
```

NDSolve::ndsz : At x == 0.9999997907551149', step size

is effectively zero; singularity or stiff system suspected. ≫

```
Out[76]= InterpolatingFunction[{{0., 1.}}, <>][x]
```

This warning message indicates that **NDSolve** was unable to generate a solution on the entire interval requested, and may occur, for example, if the solution has a vertical asymptote.

```
In[77]:= sol[2]
```

InterpolatingFunction::dmval :

Input value 2 lies outside the range of data in

the interpolating function. Extrapolation will be used. ≫

```
Out[77]= 6.42571 × 10^50
```

This second message occurred when we tried to evaluate the solution from **NDSolve** outside of the interval on which the solution is defined. In this case, the solution is not defined past 1. Nevertheless, *Mathematica* extrapolates the solution and returns a value. This can be very misleading (especially since the actual solution blows up at 1), which is why *Mathematica* issues a warning.

As you become comfortable using *Mathematica*, you may occasionally want to use the **Off** command to turn off certain warning messages. For example, you can turn off the warning message "NDSolve::ndsz" by typing **Off[NDSolve::ndsz]**. To turn it on again, enter **On[NDSolve::ndsz]**. Alternatively, you can use the **Quiet** command to "quietly" evaluate an expression, without outputting any messages. For example, **Quiet[sol[2]]** will evaluate **sol** at 2 without generating the "InterpolatingFunction::dmval" warning.

8.13.3 Recovering from Errors

From time to time you may experience certain problems using *Mathematica*. This is an unavoidable aspect of any complicated software product. We have tried to alert you to the most common and frustrating problems. More suggestions appear in the *Online Help* section of Chapter 2; *Aborting Calculations* in Chapter 3; and *Quitting the Kernel* in Chapter 4. If *Mathematica* is hung up and won't respond, type ALT+. to abort the calculation. If all else fails, terminate the *Mathematica* process from your operating system or reboot the computer. Above all, remember to save your work often.

Problem Set C

Numerical Solutions

In this problem set you will use **NDSolve** and **Plot** to calculate and plot numerical solutions to differential equations. These commands are explained in Chapters 3, 7, and 8. The solution to Problem 3 appears in the *Sample Solutions*.

1. We are interested in describing the solution $y = \phi(t)$ to the initial value problem

$$\frac{dy}{dt} = 2\frac{t}{y} + e^y, \qquad y(0) = 1. \tag{C.1}$$

Observe that $dy/dt \geq e^y \geq e$ in the region $t \geq 0$, $y \geq 1$, and therefore $\phi(t)$ increases to ∞ as t increases. But how fast does it increase?

(a) First of all, $\phi(t)$ must increase at least as fast as the solution $y = \phi_0(t)$ to the initial value problem

$$\frac{dy}{dt} = e^y, \qquad y(0) = 1.$$

Solve this problem symbolically and conclude that $\phi(t) \to \infty$ as $t \to t^*$ for some $t^* \leq 1/e < 1/2$.

(b) Next, since $2t/y \leq 1$ for $t \leq 1/2$ and $y \geq 1$, it follows that $\phi(t)$ increases at most as fast as the solution $y = \phi_1(t)$ to the initial value problem

$$\frac{dy}{dt} = 1 + e^y, \qquad y(0) = 1.$$

Solve this problem symbolically and conclude that $t^* \geq \ln(e + 1) - 1$.

(c) Now compute a numerical solution of (C.1) and find an approximate value of t^*.

(d) Finally, plot the numerical solution on the same graph with $\phi_0(t)$ and $\phi_1(t)$ and compare the solutions.

2. Consider the initial value problem

$$y' = y^2 - e^{-t^2}, \qquad y(0) = 2. \qquad \text{(C.2)}$$

Since **DSolve** is unable to solve this equation, we need to use qualitative and numerical methods. Observe that $0 < e^{-t^2} \leq 1$, so that

$$y^2 \geq y^2 - e^{-t^2} \geq y^2 - 1.$$

Therefore the solution $\phi(t)$ to (C.2) must satisfy $\phi_0(t) \geq \phi(t) \geq \phi_1(t)$, where ϕ_0 and ϕ_1 are the solutions to the respective initial value problems

$$\begin{cases} y' = y^2 & y(0) = 2, \\ y' = y^2 - 1 & y(0) = 2. \end{cases}$$

(a) Solve for ϕ_0 and ϕ_1 explicitly (using **DSolve**) and conclude that $\phi(t) \to \infty$ as $t \to t^*$, for some $t^* \in [0.5, 0.5 \ln 3]$.

(b) Compute a numerical solution of (C.2), find an approximate value of t^*, and plot $\phi(t)$ for $0 \leq t < t^*$.

3. We shall study solutions $y = \phi_b(t)$ to the initial value problem

$$y' = (y - t)(1 - y^3), \qquad y(0) = b$$

for nonnegative values of t.

(a) Plot numerical solutions $\phi_b(t)$ for several values of b. Include values of b that are less than or equal to 0, between 0 and 1, equal to 1, and greater than 1.

(b) Now, based on these plots, describe the behavior of the solution curves $\phi_b(t)$ for positive t, when $b \leq 0, 0 < b < 1, b = 1$, and $b > 1$. Identify limiting behavior and indicate where the solutions are increasing or decreasing.

(c) Next, combine your plots with a plot of the line $y = t$. The graph should suggest that the solution curves for $b > 1$ are asymptotic to this line. Explain from the differential equation why that is plausible. (*Hint:* Use the differential equation to consider the sign of y' on and close to the line $y = t$.)

(d) Finally, superimpose a plot of the direction field of the differential equation to confirm your analysis.

4. We shall study solutions $y = \phi_b(t)$ to the initial value problem

$$y' = (y - \sqrt{t})(1 - y^2), \qquad y(0) = b$$

for nonnegative values of t.

(a) Plot numerical solutions $\phi_b(t)$ for several values of b. Include values of b that are less than -1, equal to -1, between -1 and 1, equal to 1, and greater than 1.

(b) Now, based on these plots, describe the behavior of the solution curves $\phi_b(t)$ for positive t, when $b < -1$, $b = -1$, $-1 < b < 1$, $b = 1$, and $b > 1$. Identify limiting behavior and indicate where the solutions are increasing or decreasing.

(c) By combining your plots with a plot of the parabola $y = \sqrt{t}$, show that the solution curves for $b > 1$ are asymptotic to this parabola. Explain from the differential equation why that is plausible.

(d) Finally, superimpose a plot of the direction field of the differential equation to confirm your analysis.

5. In this problem, we analyze the Gompertz-threshold model from Problem 16 of Problem Set B. That is, consider the differential equation

$$y' = y(1 - \ln y)(y - 3).$$

Using various nonnegative values of $y(0)$, find and plot several numerical solutions on the interval $0 \le t \le 6$. By examining the differential equation and analyzing your plots, identify all equilibrium solutions and discuss their stability.

6. Consider the initial value problem

$$\frac{dy}{dt} = \frac{t - e^{-t}}{y + e^y}, \qquad y(1.5) = 0.5.$$

(a) Use **NDSolve** to find approximate values of the solution at $t = 0, 1, 1.8$, and 2.1. Then plot the solution.

(b) In this part you should use the results from parts (c) and (d) of Problem 5 in Problem Set B (which appears in the *Sample Solutions*). Compare the values of the actual solution and the numerical solution at the four specified points. Plot the actual solution and the numerical solution on the same graph.

(c) Now plot the numerical solution on several large intervals (e.g., $1.5 \le t \le 10$ or $1.5 \le t \le 100$). Make a guess about the nature of the solution as $t \to \infty$. Try to justify your guess on the basis of the differential equation.

7. Consider the initial value problem

$$e^y + (te^y - \sin y)\frac{dy}{dt} = 0, \qquad y(2) = 1.5.$$

(a) Use **NDSolve** to find values of the solution at $x = 1, 1.5$, and 3. Then plot the solution on the interval $0.5 \le t \le 4$.

(b) In this part you should use the results from parts (c) and (d) of Problem 6 in Problem Set B. Compare the values of the actual solution and the numerical solution at the three specified points. Plot the actual solution and the numerical solution on the same graph.

(c) Now plot the numerical solution on several large intervals (*e.g.*, $2 \le t \le 10$, or $2 \le t \le 100$, or even $2 \le t \le 1000$). Make a guess about the nature of the solution as $t \to \infty$. Try to justify your guess on the basis of the differential equation. Similarly, plot the numerical solution on intervals $\epsilon \le t \le 2$ for different choices of ϵ as a small positive number. What do you think is happening to the solution as $t \to 0+$?

8. The differential equation

$$y' = e^{-t^2}$$

cannot be solved in terms of elementary functions, but it can be solved with **DSolve**.

(a) Use **DSolve** to solve this equation.

(b) The answer is in terms of **Erf**, a special function (*cf.* Chapter 5) known as the *error function*. Differentiate the error function to see that

$$\frac{d}{dt}(\text{erf}(t)) = \frac{2}{\sqrt{\pi}}e^{-t^2}.$$

In fact,

$$\text{erf}(t) = \frac{2}{\sqrt{\pi}}\int_0^t e^{-s^2}\,ds.$$

Check the help text on **Erf** to confirm this formula.

(c) Although one does not have an elementary formula for this function, the numerical capabilities of *Mathematica* mean that we "know" this function as well as we "know" elementary functions like $\tan t$. To illustrate this, evaluate $\text{erf}(t)$ at $t = 0, 1$, and 10.5, and plot $\text{erf}(t)$ on $-10 \le t \le 10$.

(d) Compute $\lim_{t\to\infty} \text{erf}(t)$ and $\int_{-\infty}^{\infty} e^{-t^2}\,dt$.

(e) Solve the initial value problem

$$\frac{dy}{dt} = \frac{2}{\sqrt{\pi}}e^{-t^2}, \qquad y(0) = 0,$$

numerically, using **NDSolve** and the accuracy options discussed in Chapter 7, to calculate values for $\text{erf}(0.1), \text{erf}(0.2), \ldots, \text{erf}(1)$ having at least 10 correct digits. Present your results in a table. In another column print the values of $\text{erf}(t)$ for $t = 0.1, 0.2, \ldots, 1$, obtained by using the built-in function **Erf**. Compare the two columns of values.

9. Consider the initial value problem

$$ty' + (\sin t)y = 0, \qquad y(0) = 1.$$

(a) Use **DSolve** to solve the initial value problem.

(b) Note the occurrence of the built-in function **SinIntegral**, called the *sine integral* function Si(t). Check that this function is an antiderivative of $\sin t/t$.

(c) Evaluate $\lim_{t\to\infty}$ Si(t). Plot Si(t) and discuss the features of the graph.

(d) Do the same with the solution to the initial value problem.

(e) Now solve the initial value problem using **NDSolve**, and plot the computed solution. Compare your plot to the one obtained in part (d). You will find that **NDSolve** will not accept the equation in the form written above, because *Mathematica* cannot evaluate $\sin t/t$ at $t = 0$, even though the singularity is removable. The simplest way around this is to divide through by t and use the *Mathematica* function **Sinc**, which is defined to have the value 1 at 0, but which at any other value of t is defined to be $\sin t/t$.

(f) Discuss the stability of the differential equation. Illustrate your conclusions by graphing solutions with different initial values on the interval $[-10, 10]$.

Note: In this problem some of your plots may take a long time to generate.

10. Solve the following initial value problems numerically, then plot the solutions. Based on your plots, predict what happens to each solution as t increases. In particular, if there is a limiting value for y, either finite or infinite, find it. If it is unclear from the plot you've made, try replotting on a larger interval. Another possibility is that the solution blows up in finite time. If so, estimate the time when the solution blows up. Try to use the qualitative methods of Chapter 6 to confirm your answers.

(a) $y' = e^{-3t} + \dfrac{1}{1+y^2}, \quad y(0) = -1.$

(b) $y' = e^{-2t} + y^2, \quad y(0) = 1.$

(c) $y' = \cos t - y^3, \quad y(0) = 0.$

(d) $y' = (\sin t)y - y^2, \quad y(0) = 2.$

11. Solve the following initial value problems numerically, then plot the solutions. Based on your plots, predict what happens to each solution as t increases. See Problem 10 for additional instructions.

(a) $y' = 3t - 5\sqrt{y}, \quad y(0) = 4.$

(b) $y' = (t^3 - y^3)\cos y, \quad y(0) = -1.$

(c) $y' = \dfrac{y^2 + 2ty}{t^2 + 3}, \quad y(1) = 2.$

(d) $y' = -2t + e^{-ty}, \quad y(0) = 1.$

12. This problem illustrates one of the possible pitfalls of blindly applying numerical methods without paying attention to the theoretical aspects of the differential equation itself. Consider the equation

$$ty' + 3y - 9t^2 = 0.$$

(a) Use the *Mathematica* program **EulerMethod** from Chapter 7, with step size $h = 0.2$ and $n = 10$ steps, to compute an approximation to the solution $y(t)$ of this equation with initial condition $y(-0.5) = 3.15$. The program will generate a list of ordered pairs (x_i, y_i). Use **ListPlot** to graph the piecewise linear function connecting the points (x_i, y_i).

(b) Now modify the program to implement the Improved Euler Method. Can you make sense of your answers?

(c) Next, use **NDSolve** (with the default method) to find an approximate solution on the interval $(-0.5, 0.5)$, and plot it with **Plot**. Print out the values of the solution at the equally spaced points $-0.06, -0.04, \ldots, 0.04, 0.06$. What is the interval on which the approximate solution is defined?

(d) Solve the equation explicitly and graph the solutions for the initial conditions $y(0) = 0$, $y(-0.5) = 3.15$, $y(0.5) = 3.15$, $y(-0.5) = -3.45$, and $y(0.5) = -3.45$. Now explain your results in (a)–(c). Could we have known, without solving the equation, whether to expect meaningful results in parts (a) and (b)? Why? Can you explain how **NDSolve** avoids making the same mistake?

13. Consider the initial value problem

$$\frac{dy}{dt} = e^{-t} - 3y, \qquad y(-1) = 0.$$

(a) Use the **EulerMethod** program from Chapter 7, with step size $h = 0.5$ and $n = 4$ steps, to compute an approximation to the solution $y(t)$ of the initial value problem. The program will generate a list of ordered pairs (t_i, y_i). Use **ListPlot** to graph the piecewise linear function connecting the points (t_i, y_i). Repeat with $h = 0.2$ and $n = 10$.

(b) Now modify the program to implement the Improved Euler Method, and repeat part (a) using the modified program. Find the exact solution of the initial value problem and plot it, the two Euler approximations, and the two Improved Euler approximations on the same graph. Label the five curves.

(c) Now use the **EulerMethod** program with $h = 0.5$ to approximate the solution on the interval $[-1, 9]$. Plot both the approximate and exact solutions on this interval. How close is the approximation to the exact solution as t increases? In light of the discussion of stability in Chapters 5 and 7, explain your results in parts (a)–(c).

14. Consider the initial value problem

$$\frac{dy}{dt} = 2y + \cos t, \qquad y(0) = -2/5.$$

(a) Use the **EulerMethod** program from Chapter 7, with step size $h = 0.5$ and $n = 12$ steps, to compute an approximation to the solution $y(t)$ of the initial value problem. The program will generate a list of ordered pairs (t_i, y_i). You do not need to print out these numbers, but graph the piecewise linear function connecting the points (t_i, y_i). What appears to be happening to y as t increases?

(b) Repeat part (a) with $h = 0.2$ and $n = 30$, and then with $h = 0.1$ and $n = 60$, each time plotting the results on the same set of axes as before. How are the approximate solutions changing as the step size decreases? Can you make a reliable prediction about the long-term behavior of the solution?

(c) Use **NDSolve** (with the default method) to find an approximate solution and again plot it on the interval $[0, 6]$ on the same set of axes as before. Now what does it look like y is doing as t increases? Next, plot the solution from **NDSolve** on a larger interval (going to $t = 15$) on a new set of axes. Again, what is happening to y as t increases?

(d) Solve the initial value problem exactly and compare the exact solution to the approximations found above. In light of the discussion of stability in Chapters 5 and 7, explain your results in parts (a)–(c).

15. Consider the initial value problem

$$\frac{dy}{dt} = 2y - 2 + 3e^{-t}, \qquad y(0) = 0.$$

(a) Use the **EulerMethod** program from Chapter 7, with step size $h = 0.2$ and $n = 10$ steps, to compute an approximation to the solution $y(t)$ of the initial value problem. The program will generate a list of ordered pairs (t_i, y_i). Use **ListPlot** to graph the piecewise linear function connecting the points (t_i, y_i). What appears to be happening to y as t increases?

(b) Repeat part (a) with $h = 0.1$ and $n = 20$, and then with $h = 0.05$ and $n = 40$. How are the approximate solutions changing as the step size decreases? Can you make a reliable prediction about the long-term behavior of the solution?

(c) Use **NDSolve** (with the default method) to find an approximate solution and plot it on the interval $[0, 2]$. Now what does it look like y is doing as t increases? Next, plot the solution from **NDSolve** on a larger interval (going at least to $t = 10$). Again, what is happening to y as t increases?

(d) Solve the initial value problem exactly and compare the exact solution to the approximations found above. In light of the discussion of stability in Chapters 5 and 7, explain your results in parts (a)–(c).

16. Consider the Gompertz-threshold model,

$$y' = y(1 - \log y)(y - 3).$$

From the direction field and the qualitative approach for this equation, one learns that the solution with initial condition, $y(0) = 5$, approaches 3 as t approaches ∞. Use the event detection feature of **NDSolve** to find the time t at which $y = 3.1$.

Chapter 9

Solving and Analyzing Second Order Linear Equations

Newton's second law of dynamics—force is equal to mass times acceleration—tells physicists that, in order to understand how the world works, they must pay attention to forces. Since acceleration is a second derivative, the law also tells us that second order differential equations are likely to appear when we apply mathematics to study the real world.

We note that in Chapters 5, 6 and 7 we discussed three different approaches to solving first order differential equations: searching for exact formula solutions; using geometric methods to study qualitative properties of solutions—typically when we cannot find a formula solution; and invoking numerical methods to produce approximate solutions—again when no solution formula is attainable.

In this chapter we shall bring each of these methods to bear on second order equations. The most basic second order differential equations are linear equations with *constant coefficients*:

$$ay'' + by' + cy = g(t).$$

These equations model a wide variety of physical situations, including oscillations of springs, simple electric circuits, and the vibrations of tuning forks to produce sound and of electrons to produce light. In other situations, such as the motion of a pendulum, we may be able to approximate the resulting differential equation reasonably well by a linear differential equation with constant coefficients. Fortunately, there are several techniques for finding explicit solution formulas to linear differential equations with constant coefficients, which *Mathematica* can apply.

Unfortunately, we often cannot find solution formulas for more general second order equations, even for linear equations with *variable coefficients*:

$$y'' + p(t)y' + q(t)y = g(t).$$

Such equations have important applications to physics. For example, Airy's equation,

$$y'' - ty = 0, \tag{9.1}$$

arises in diffraction problems in optics, and Bessel's equation,

$$y'' + \frac{1}{t}y' + y = 0, \qquad (9.2)$$

occurs in the study of vibrations of a circular membrane and of water waves with circular symmetry.

When studying first order differential equations for which exact solution formulas were unavailable, we could turn to several other methods. By specifying an initial value, we could compute an approximate numerical solution. By letting the initial values vary, we could plot a one-parameter family of approximate solution curves and get a feel for the behavior of a general solution. Alternatively, we could plot the direction field of the differential equation and use it to draw conclusions about the qualitative behavior of solutions.

For second order equations, these methods are more cumbersome; you must specify two conditions to pick out a unique solution. We usually specify initial values for the function and its first derivative at some point. Then we can use numerical techniques to compute an approximate solution. In order to graph enough solutions to get an idea of their general behavior, we must construct a two-parameter family of solutions. Since the initial value does not determine the initial slope, we cannot draw a direction field for a second order equation.

Given a second order differential equations, we are often interested in solving an associated *initial value problem*; *i.e.*, finding the solution of the differential equation satisfying the *initial conditions* $y(t_0) = y_0$ and $y'(t_0) = y'_0$. In some other situations, the differential equation is accompanied not by initial conditions, but by *boundary conditions*; *i.e.*, we specify values $y(t_0) = y_0$ and $y(t_1) = y_1$ at two distinct points. The resulting problem is called an *boundary value problem*. See Section 9.2 for further information.

In this chapter, we describe how to use *Mathematica* to find formula solutions of second order linear differential equations with constant coefficients. We also describe how to find and plot numerical solutions to more general second order differential equations. In addition, we describe a method for solving boundary value problems using *Mathematica*'s numerical solver. Finally, we introduce comparison methods and a more sophisticated geometric method for analyzing second order linear equations with variable coefficients. These qualitative methods have an advantage over the more obvious numerical and graphical methods. They more effectively yield information about properties shared by all solutions of a differential equation, about the oscillatory nature of solutions of an equation, and about the precise rate of decay or growth of solutions.

Comparison methods provide information on the solutions of a variable coefficient equation by comparing the equation with an appropriate constant coefficient equation. This possibility is suggested by the following example. For large t, the coefficient $1/t$ in Bessel's equation (9.2) is close to 0. So, Bessel's equation is close to the equation $y'' + y = 0$, whose general solution can be written as $y = R\cos(t - \delta)$. Thus one might expect solutions to Bessel's equation to oscillate and look roughly like sine waves for large t. We present a result that validates such comparisons.

The geometric method is based on direction fields, which are not directly applicable to a second order equation. We show, however, how to construct a related first order equation whose direction field yields information about the solutions of the original second order equation.

9.1 Second Order Equations with *Mathematica*

Mathematica's usual tools for finding symbolic solutions of differential equations work perfectly well for second order linear differential equations with constant coefficients. The syntax of the command is familiar, except that we must specify an additional initial condition.

Example 9.1 Consider the differential equation

$$y'' + y' - 6y = 20e^t. \tag{9.3}$$

We can get a solution by typing:

```
ode1 = y''[t] + y'[t] - 6 y[t] == 20 Exp[t];
DSolve[ode1, y[t], t]
```

$$\{\{y[t] \rightarrow -5 e^t + e^{-3t} C[1] + e^{2t} C[2]\}\}$$

As expected, the general solution depends on two arbitrary constants. To solve the differential equation with initial conditions $y(0) = 0$ and $y'(0) = 1$, we would type

```
DSolve[{ode1, y[0] == 0, y'[0] == 1}, y[t], t]
```

$$\left\{\left\{y[t] \rightarrow \frac{1}{5} e^{-3t} \left(4 - 25 e^{4t} + 21 e^{5t}\right)\right\}\right\}$$

Example 9.2 Now consider the linear second order differential equation

$$y'' + t^2 y' + y = 0. \tag{9.4}$$

Mathematica is unable to find an explicit formula for the general solution. Instead, we can use **NDSolve** to find a numerical solution. **NDSolve** generates a numerical solution in terms of an **InterpolatingFunction**. An **InterpolatingFunction** represents an approximate function that, when applied to a particular t, returns the approximate value of $y(t)$ at that point. The numerical solution defined on the interval $[0, 3]$ with initial conditions $y(0) = 1$, $y'(0) = -1$ can be found by typing

```
ode2 = y''[t] + t^2*y'[t] + y[t] == 0;
nsol1 = NDSolve[{ode2, y[0] == 1, y'[0] == -1}, y[t], {t, 0, 3}]
```

$$\{\{y[t] \rightarrow \text{InterpolatingFunction}[\{\{0., 3.\}\}, <>], y[t]\}\}$$

and it can be plotted using the techniques developed in Chapter 7.

A good way to understand the behavior of a general solution of a second order equation is to plot numerical solutions corresponding to a wide range of initial values and initial slopes. Here are the *Mathematica* commands to do so for this example.

```
nsol2 := NDSolve[{ode2, y[0] == y0, y'[0] == yp0}, y[t],
    {t, -2, 3}];
nsol2func[t_, y0_, yp0_] := y[t] /. First[nsol2]
Plot[Evaluate[Table[nsol2func[t, y0, yp0], {y0, -2, 2},
    {yp0, -1, 1, 0.5}]], {t, -2, 3}, PlotRange -> {-10, 10},
    AxesOrigin -> {-2, -10}]
```

The result is shown in Figure 9.1. We see that the solutions seem to level off for positive t and seem to blow up for negative t.

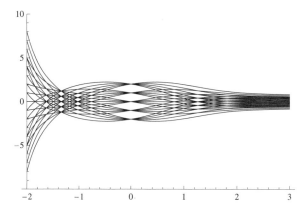

Figure 9.1: Numerical Solutions of (9.4)

Could we have obtained this qualitative information directly from the differential equation instead of from the graph of a few solutions? One approach is to look at the solutions of differential equations that are similar to this one. For example, when t is small, t^2 is close to zero. So the solutions to (9.4) should be close to the solutions of the differential equation

$$y'' + y = 0.$$

We know that the solutions to the latter equation are sine curves. For example, the solution satisfying the initial conditions $y(0) = 0$, $y'(0) = 1$ is just $\sin(t)$. So, one might expect the solutions of this initial value problem to look like $\sin(t)$ for small values of t. In Figure 9.2, we have plotted $\sin(t)$ (with a dashed line) and the numerical solution to the initial value problem (with a solid line).

The graph of the numerical solution in Figure 9.2 appears to level off as t increases. Looking again at the differential equation, we see that when t is large, the $t^2 y'$ term should dominate the y term. So, the solutions to the differential equation should be close to those

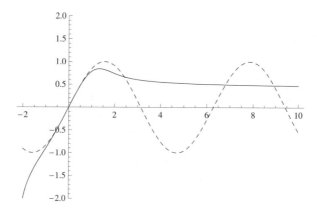

Figure 9.2: Comparing (9.4) with $y'' + y = 0$

of

$$y'' + t^2 y' = 0.$$

This equation can almost be solved explicitly; its general solution is given by

$$y(t) = A + B \int_0^t e^{-u^3/3} \, du.$$

The exponential function being integrated is nearly zero for large positive values of u. So, when t is large and positive, the solutions should become almost constant. When t is large and negative, the exponential term is huge, and we expect the solutions to blow up.

9.2   Boundary Value Problems

In some cases we can find a formula for the general solution of the differential equation. This is the case for (9.3); the general solution has two arbitrary constants, $C[1]$ and $C[2]$, which can be selected so that the general solution satisfies the given boundary conditions. If, together with (9.3) we consider the boundary conditions $y(0) = 0$ and $y(\ln(2)) = 10$, we get the solution of the boundary value problem by typing:

```
DSolve[{ode1, y[0] == 0, y[Log[2]] == 10}, y[t], t]
```

$$\{\{y[t] \rightarrow 5 e^t (-1 + e^t)\}\}$$

Example 9.3 Next, consider the nonlinear second order boundary value problem

$$y'' + \frac{1}{4} x^4 y' + y + x y^3 = 0, \ y(0) = 0, \ y(1) = 1. \tag{9.5}$$

Since **DSolve** cannot find the general solution of this differential equation, we turn to another method, the *shooting method*, which we now describe. Let y be the solution of

$$y'' + \frac{1}{4}x^4 y' + y + xy^3 = 0, \ y(0) = 0, \ y'(0) = s, \tag{9.6}$$

where the differential equation is the equation in (9.5), the first initial condition is the same as the first boundary condition in (9.5), and the second initial condition is specified by the parameter s. Since for any value of s, the solution of (9.6) satisfies the differential equation and the first boundary condition in (9.5), if we can find a value $s = \bar{s}$ of the parameter for which the solution satisfies the second boundary condition in (9.5), then the solution of (9.6) with $s = \bar{s}$ is the solution of our boundary value problem. Toward this end, the following *Mathematica* program is useful:

```
ode3 = y''[x] + 1/4 x^4 y'[x] + y[x] + y[x]^3 == 0;
nsol3[s_] := NDSolve[{ode3, y[0] == 0, y'[0] == s},
    y[x], {x, 0, 1}]
trial[s_] := nsol3[s] /. x → 1
```

Now when we evaluate **nsol3** at some real number, we get a function that can be used to compute numerical solutions to the differential equation. When we evaluate this procedure at $x = 1$, as is done by the **trial** command, we find the value of that numerical solution at the second endpoint in our boundary value problem. After some trial and error we find that **trial[1.2526]** produces the output

```
{{y[1] → 0.999975}}
```

So the solution of our boundary value problem is approximately the solution of (9.6) with the initial conditions $y(0) = 0$ and $y'(0) = 1.2526$.

If our boundary value problem is linear, *i.e.*, is of the form

$$L[y] = y'' + p_1(x)y' + p_2(x)y = f, \ y(0) = y_l, y(1) = y_r,$$

where the subscripts l and r indicate the function values at the left endpoints and at the right endpoints, then the function **trial[s]** is a linear function of s, and the determination of $\bar{s}$ is more explicit. Suppose $y_0, y_1,$ and y_2 are solutions of the following the second order initial value problems, respectively:

$$L[y_0] = f, \ y_0(0) = 0, \ y_0'(0) = 0;$$
$$L[y_1] = 0, \ y_1(0) = y_l, \ y_1'(0) = 0;$$

$$L[y_2] = 0, \ y_2(0) = 0, \ y_2'(0) = 1.$$

Then $y(x) = y_0(x) + y_1(x) + sy_2(x)$ is the solution of (9.6) for any value of s. Thus

$$\bar{s} = \frac{y_r - y_0(1) - y_1(1)}{y_2(1)},$$

and the desired solution of our boundary value problem is

$$y(x) = y_0(x) + y_1(x) + \frac{y_r - y_0(1) - y_1(1)}{y_2(1)} y_2(x).$$

Example 9.4 Finally, we return to the boundary value problem (9.5) in Example 9.3, but this time solve it with *Mathematica*'s built-in boundary value problem solver. To find and plot the (numerical) solution, type:

```
nsol4 = NDSolve[{ode3, y[0] == 0, y[1] == 1}, y[x],
    {x, 0, 1}];
Plot[y[x] /. First[nsol4], {x, 0, 1}]
```

The result is shown in Figure 9.3.

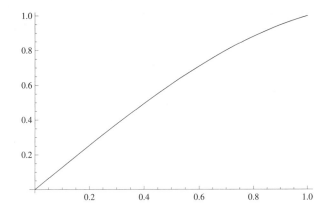

Figure 9.3: Numerical Solution of Boundary Value Problem (9.5)

Here **NDSolve** knows to treat this problem as a boundary value problem because of the presence of the two boundary conditions, instead of the two initial conditions in a second order initial value problem. The algorithm used is a variant of the shooting method.

9.3   Comparison Methods

In this section, we discuss the Sturm Comparison Theorem. In addition, we will discuss the relation between the zeros of two linearly independent solutions of a second order linear equation.

Theorem 9.1 (Sturm Comparison Theorem) *Let $u(x)$ be a solution to the equation*

$$y'' + q(x)y = 0,$$

and suppose $u(a) = u(b) = 0$ *for some* $a < b$ *(but* u *is not identically zero). Let* $v(x)$ *be a solution to the equation*

$$y'' + r(x)y = 0,$$

where $r(x) \geq q(x)$ *for all* $a \leq x \leq b$. *Then* $v(x) = 0$ *for some* x *in* $[a, b]$.

We will defer the proof of this theorem briefly, preferring instead to explain how to use it. We typically use this result to compare a variable coefficient equation with an appropriate constant coefficient equation. Consider, for example, the equation

$$y'' + \frac{1}{x}y = 0. \tag{9.7}$$

Let K be a positive number. If $0 < x \leq K$, then $1/x \geq 1/K$. We now apply the above theorem with $q(x) = 1/K$ and $r(x) = 1/x$. Thus we compare equation (9.7) to the constant coefficient equation $y'' + (1/K)y = 0$, whose general solution is $u(x) = R\cos(\sqrt{1/K}\,x - \delta)$. Given any interval in $(0, K]$ of length $\pi\sqrt{K}$, the value of δ can be chosen so that $u(x) = 0$ at the endpoints of the interval. Then the Sturm Comparison Theorem implies that every solution of (9.7) has a zero on this interval. This argument implies that on $(0, K]$, the zeros of (9.7) cannot be farther apart than $\pi\sqrt{K}$.

In particular, solutions of (9.7) must oscillate as $x \to \infty$, though the oscillations may become less frequent as x increases. Indeed, by turning the above comparison around, one concludes that the zeros of solutions of (9.7) in $[K, \infty)$ must be at least $\pi\sqrt{K}$ apart.

Figure 9.4 shows two representative solutions of (9.7) computed with **DSolve**. As you can see, the graph confirms the predictions of the previous paragraph.

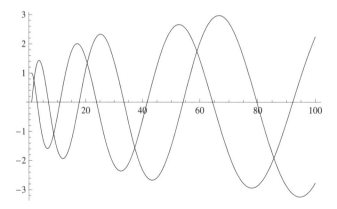

Figure 9.4: Solutions of $y'' + \frac{1}{x}y = 0$

The Sturm Comparison Theorem can also be used to study the zeros of solutions to Airy's equation and to Bessel's equation. In the latter case, a substitution is made to eliminate the y' term. See Problems 18 and 16 in Problem Set D.

9.3.1 The Interlacing of Zeros

You may have noticed that the curves in Figure 9.1 take turns crossing the horizontal axis. We say that the zeros of the two solutions are *interlaced*: between any two zeros of one solution there is a zero of the other. This is a common phenomenon—the property of interlaced zeros holds for any pair of linearly independent solutions of a homogeneous second order linear differential equation

$$y'' + p(x)y' + q(x)y = 0. \tag{9.8}$$

If $p(x) = 0$, as is the case in equation (9.1), the proof of the interlacing result follows from an application of the Sturm Comparison Theorem by comparing the equation with itself. For the more general equation (9.8), the proof is based on the Wronskian. Let $y_1(x)$ and $y_2(x)$ be any two linearly independent solutions of (9.8). Recall that the Wronskian of $y_1(x)$ and $y_2(x)$ is

$$W(x) = y_1(x)y_2'(x) - y_1'(x)y_2(x).$$

Since y_1 and y_2 are solutions of (9.8), it follows that W satisfies the differential equation

$$W' = -p(x)W,$$

whose general solution is

$$W(x) = Ce^{-\int p(x)dx}.$$

This formula shows that $W(x)$ does not change sign—it always has the same sign as the constant C, which must be nonzero (or else y_1 and y_2 would be linearly dependent).

Assume that $y_1(a) = y_1(b) = 0$ and that there are no zeros of y_1 between a and b. Then

$$W(a) = -y_1'(a)y_2(a),$$
$$W(b) = -y_1'(b)y_2(b).$$

Since y_1 does not change sign between a and b, we know $y_1'(a)$ and $y_1'(b)$ have opposite signs. Then, since $W(a)$ and $W(b)$ have the same sign, $y_2(a)$ and $y_2(b)$ must have opposite signs. Therefore y_2 must be zero somewhere between a and b. Similarly, y_1 must have a zero between any two zeros of y_2.

Exercise 9.1 The solutions to Airy's equation and Bessel's equation are considered so important they have been given names. The Airy functions $Ai(x)$ and $Bi(x)$ are a fundamental set of solutions for (9.1), and the Bessel functions $J_0(x)$ and $Y_0(x)$ are a fundamental set of solutions for (9.2). These functions are built into *Mathematica* as `AiryAi[x]`, `AiryBi[x]`, `BesselJ[0,x]`, and `BesselY[0,x]`. Use *Mathematica* to check that the zeros of the two solutions `AiryAi[x]` and `AiryBi[x]` to Airy's equation are interlaced. Do the same for the Bessel functions `BesselJ[0,x]` and `BesselY[0,x]`.

Remark 9.1 The Airy functions and the Bessel functions, along with the gamma function, the error function, and the sine integral function, are examples of *special functions*, which were introduced in Chapter 5.

9.3.2 Proof of the Sturm Comparison Theorem

The proof is based on the Wronskian, as was the proof of the interlacing result. Recall the hypotheses: $u'' + q(x)u = 0$ and $v'' + r(x)v = 0$, with $q(x) \leq r(x)$ between two zeros a and b of u. Since u is not identically zero, we can assume u does not change sign between a and b. (If it did it would have a zero between a and b, and we could look at a smaller interval.) Assume, for instance, that $u > 0$ between a and b. Since $u = 0$ at a and b, it follows that $u'(a) > 0$ and $u'(b) < 0$. (Notice that these derivatives cannot be zero because, by the uniqueness theorem, u would then be identically zero.) We want to prove that v is zero somewhere in $[a, b]$; we suppose it is not, say $v(x) > 0$ throughout $[a, b]$, and argue to obtain a contradiction.

Let $W(x)$ be the Wronskian of u and v,

$$W(x) = u(x)v'(x) - u'(x)v(x);$$

then

$$W(a) = -u'(a)v(a) < 0,$$
$$W(b) = -u'(b)v(b) > 0.$$

Also, since $q(x) \leq r(x)$,

$$W' = uv'' - u''v = u(-r(x)v) - (-q(x)u)v = (q(x) - r(x))uv \leq 0.$$

But this is impossible—the last inequality implies W does not increase between a and b, yet $W(a) < 0 < W(b)$. This contradiction means that v cannot have the same sign throughout $[a, b]$, and therefore v must be zero somewhere in the interval.

9.4   A Geometric Method

As mentioned above, direction fields are not directly applicable to second order equations. We now show, however, that with a given second order homogeneous equation, we can associate a first order equation whose direction field yields information about the solutions of the second order equation. (Our approach is similar to the classical method of associating a first order Riccati equation to a second order linear equation. The substitution we use is akin to the Prüfer substitution for Sturm-Liouville systems; see G. Birkhoff and G.-C. Rota, *Ordinary Differential Equations*, 3rd ed., J. Wiley and Sons, Inc., 1978.)

Consider the homogeneous equation

$$y'' + p(x)y' + q(x)y = 0. \tag{9.9}$$

Let

$$z = \arctan\left(\frac{y}{y'}\right);$$

then

$$z' = \left(1 + \left(\frac{y}{y'}\right)^2\right)^{-1} \frac{d}{dx}\left(\frac{y}{y'}\right) = \frac{y'^2}{y'^2 + y^2} \frac{y'^2 - yy''}{y'^2} = \frac{y'^2 - yy''}{y'^2 + y^2}.$$

Also, since $\tan z = y/y'$, notice that

$$\sin z = \frac{y}{\sqrt{y'^2 + y^2}}, \quad \cos z = \frac{y'}{\sqrt{y'^2 + y^2}}.$$

Then if y satisfies (9.9) and y is not identically zero, it follows that

$$z' = \frac{y'^2 - yy''}{y'^2 + y^2} = \frac{y'^2 - y(-p(x)y' - q(x)y)}{y'^2 + y^2} = \frac{y'^2 + p(x)yy' + q(x)y^2}{y'^2 + y^2}.$$

In other words,

$$z' = \cos^2 z + p(x)\sin z \cos z + q(x)\sin^2 z. \qquad (9.10)$$

We have shown that for every solution y of (9.9) that is not identically zero, there is a corresponding solution z of (9.10). (This is not a one-to-one correspondence—every constant multiple of a solution y corresponds to the same solution z.) Although the solution curves of (9.10) will be different from the solution curves of (9.9), we now show that we can learn about the solutions of (9.9) by studying the solutions of (9.10), specifically, by considering the direction field of (9.10).

9.4.1   The Constant Coefficient Case

We begin by considering the equation

$$y'' - y = 0,$$

whose general solution is

$$y = c_1 e^x + c_2 e^{-x}.$$

The corresponding first order equation is

$$z' = \cos^2 z - \sin^2 z.$$

In Figure 9.5, we show the direction field of this equation, which we produced with the *Mathematica* commands:

```
<< VectorFieldPlots`
VectorFieldPlot[{1, Cos[z]^2 - Sin[z]^2}, {x, 0, 10},
    {z, -π/2, π/2}, PlotPoints → 20, ScaleFunction → (1&),
    Axes → True, Frame → True, AspectRatio → 1]
```

Notice that we plot z from $-\pi/2$ to $\pi/2$; this is the range of values taken on by the arctan function. You may wonder what happens if z goes off the bottom of the graph. The answer is that z "wraps around" to the top of the graph. Recall that $z = \arctan(y/y')$; the points $z = \pm\pi/2$ correspond to $y' = 0$, and when y' changes sign, then so does z by passing from $-\pi/2$ to $\pi/2$ (or vice versa).

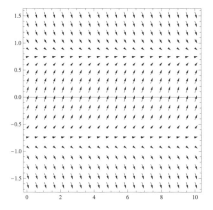

Figure 9.5: Direction Field of $z' = \cos^2 z - \sin^2 z$

Next, observe that there is an unstable equilibrium at $z = -\pi/4$, which corresponds to $y/y' = \tan(-\pi/4) = -1$; this represents solutions of the form $y = ce^{-x}$. Similarly, the stable equilibrium $z = \pi/4$ corresponds to $y/y' = \tan(\pi/4) = 1$, which represents the solutions $y = ce^x$. The fact that most solutions of the equation satisfied by z (those with initial condition other than $-\pi/4$) approach the stable equilibrium $z = \pi/4$ corresponds to the fact that most solutions $y = c_1 e^x + c_2 e^{-x}$ (those with $c_1 \neq 0$) grow like e^x as x increases.

Another basic example is

$$y'' + y = 0,$$

for which the corresponding first order equation is

$$z' = \cos^2 z + \sin^2 z = 1.$$

That is, z simply increases linearly, although every time it reaches $\pi/2$ it wraps around to $-\pi/2$. This corresponds to the oscillation of solutions y; for every time $z = \arctan(y/y')$ passes through zero, then so must y, and vice versa.

Exercise 9.2 Consider the general second order linear homogeneous equation with constant coefficients:

$$ay'' + by' + c = 0.$$

Investigate how the roots of the characteristic equation, if real, correspond to equilibrium solutions for z. Show that if the roots of the characteristic equation are complex, then z' is always positive.

9.4.2   The Variable Coefficient Case

The examples above suggest ways to draw parallels between solutions z of (9.10) and solutions y of (9.9). First of all, if z is positive, then y and y' have the same sign. This implies that y is moving away from zero as x increases. We say in this case that y is *growing*, meaning that $|y|$ is increasing. Similarly, if z is negative then y and y' have opposite signs, which implies that y is moving toward zero as x increases—we say in this case that y is *decaying*, meaning that $|y|$ is decreasing. If z changes sign by passing through zero, then so does y, and if z passes through $\pm\pi/2$, then y' changes sign, showing that y has passed through a local maximum or minimum.

In light of these observations, let us summarize what we can predict about the *long-term behavior* of a solution y of (9.9) in terms of the corresponding solution z of (9.10).

- If z remains positive as x increases, then y grows away from zero.

- If z remains negative as x increases, then y decays toward zero.

- If z continues to increase from $-\pi/2$ to $\pi/2$ and wraps around to $-\pi/2$ again, then y *oscillates* as x increases.

Notice from (9.10) that $z' = 1$ whenever $z = 0$, so it is not possible for z to decrease through zero. Thus the long-term behavior of z will usually fall into one of the three categories above.

We can be more precise about the growth or decay rate of y in cases where z approaches a limiting value. If the limiting value is θ, then y/y' approaches $\tan\theta$ as x increases. In other words, $y' \approx (\cot\theta)y$ for large x. Thus

$$y \approx ce^{(\cot\theta)x}$$

for large x, and $\cot\theta$ is the asymptotic exponential growth (or decay) rate of y. If z approaches zero as x increases, then y/y' approaches zero as well. One can show that y grows (if $z > 0$) or decays (if $z < 0$) faster than any exponential function. Similarly, if z approaches $\pi/2$, then y grows slower than exponentially, and if z approaches $-\pi/2$, then y decays slower than exponentially. Since y'/y approaches zero in these cases, y could grow or decay toward a finite, nonzero value.

Let us now see what the direction field of (9.10) tells us about solutions of Airy's and Bessel's equations.

9.4.3 Airy's Equation

For Airy's equation,

$$y'' - xy = 0, \tag{9.11}$$

the corresponding first order equation is

$$z' = \cos^2 z - x\sin^2 z.$$

Figure 9.6 shows the direction field of this equation, plotted by a *Mathematica* command similar to the one used for Figure 9.5.

Notice that z is increasing steadily for negative x, so solutions of (9.11) must oscillate for negative x. For positive x, it is evident that solutions of (9.11) cannot oscillate; z passes

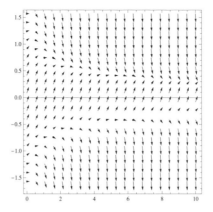

Figure 9.6: Direction Field of $z' = \cos^2 z - x \sin^2 z$

through zero at most once, after which it remains positive. Hence the corresponding solution of (9.11) grows away from zero as x increases. It appears that once z becomes positive, it decreases to zero, indicating that most solutions of (9.11) grow faster than exponentially.

We now know there are solutions of (9.11) that grow very fast as x increases, and there are no oscillating solutions for positive x. Is it possible to have a decaying solution of (9.11)? Equivalently, is it possible for z to remain negative as x increases? Figure 9.6 strongly suggests that there is such a solution z, and hence that there is a solution y to (9.11) that decays toward zero at a rate faster than exponential.

The Airy function $\text{Ai}(x)$ is, by definition, the unique (up to a constant multiple) solution of (9.11) that decays as x increases. As expected, the other Airy function $\text{Bi}(x)$ grows as x increases. You can see what these functions look like by plotting **AiryAi[x]** and **AiryBi[x]** in *Mathematica*. The general solution of (9.11) is a linear combination of $\text{Ai}(x)$ and $\text{Bi}(x)$.

9.4.4   Bessel's Equation

For Bessel's equation

$$y'' + \frac{1}{x}y' + y = 0, \tag{9.12}$$

the corresponding first order equation is

$$z' = \cos^2 z + \frac{1}{x} \sin z \cos z + \sin^2 z.$$

Figure 9.7 shows the direction field of this equation.

The picture is simpler than the one for Airy's equation. When x is away from zero, z is increasing steadily. So, solutions of (9.12) oscillate for both positive and negative x. Near $x = 0$, the direction field is irregular because of the $1/x$ term in the differential equation.

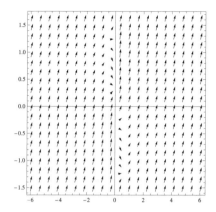

Figure 9.7: Direction Field of $z' = \cos^2 z + \frac{1}{x} \sin z \cos z + \sin^2 z$

We can't tell what happens from this picture, but there is reason to expect that solutions of (9.12) will not behave nicely near this singularity in the differential equation.

One question this approach cannot answer easily is whether the amplitude of oscillating solutions grows, decays, or remains steady as x increases. Since the coefficient of the y' term in (9.12) is positive, we can expect solutions of (9.12) to behave like a physical oscillator with damping—the amplitude of oscillations should decrease as x increases. Since the damping coefficient goes to zero as x increases, we cannot be sure (without a more refined analysis) whether the amplitude of solutions of (9.12) must decrease to zero. In fact it does, as you can check by plotting the Bessel functions `BesselJ[0,x]` and `BesselY[0,x]` in *Mathematica*.

9.4.5   Other Equations

Here are some other differential equations for which you can try this method.

- Bessel's equation of order n,

$$y'' + \frac{1}{x}y' + \left(1 - \frac{n^2}{x^2}\right) y = 0,$$

  for various n (above we studied the case $n = 0$). You can check your predictions using the functions `BesselJ[n,x]` and `BesselY[n,x]`, which are built into *Mathematica*. These functions are considered in Problem 19 of Problem Set D.

- The modified Bessel equation of order n,

$$y'' + \frac{1}{x}y' - \left(1 + \frac{n^2}{x^2}\right) y = 0,$$

for various n. The functions **BesselI[n,x]** and **BesselK[n,x]** are a fundamental set of solutions for this equation.

- The parabolic cylinder equation

$$y'' + \left(n + \frac{1}{2} - \frac{1}{4}x^2 \right) y = 0$$

for various n. This equation arises in quantum mechanics. The parabolic cylinder
functions $D_n(x)$ and $D_{-n-1}(ix)$ are a fundamental set of solutions for this equation.
These functions are built into *Mathematica* as **ParabolicCylinderD[n,x]**
and **ParabolicCylinderD[-n-1,I*x]**. This equation is considered in Problem 17 of Problem Set D.

Problem Set D

Second Order Equations

The solution to Problem 1 appears in the *Sample Solutions*.

1. Airy's equation is the linear second order homogeneous equation $y'' = ty$. Although it arises in a number of applications, including quantum mechanics, optics, and waves, it cannot be solved exactly by the standard symbolic methods. In order to analyze the solution curves, let us reason as follows:

   (a) For t close to zero, the equation resembles $y'' = 0$, which has general solution $y = c_1 t + c_2$. We refer to this as a "facsimile" solution. Graph a numerical solution to Airy's equation with initial conditions $y(0) = 0$, $y'(0) = 1$, and the facsimile solution (with the same initial data) on the interval $(-2, 2)$. How well do they match?

   (b) For $t \approx -K^2 \ll 0$, the equation resembles $y'' = -K^2 y$, and the corresponding facsimile solution is given by $y = c_1 \sin(Kt + c_2)$. Again using the initial conditions $y(0) = 0$, $y'(0) = 1$, plot a numerical solution of Airy's equation over the interval $(-18, -14)$. Using the value $K = 4$, try to find values of c_1 and c_2 so that the facsimile solution matches well with the actual solution. Why shouldn't we expect the initial conditions for Airy's equation to be the appropriate initial conditions for the facsimile solution?

   (c) For $t \approx K^2 \gg 0$, Airy's equation resembles $y'' = K^2 y$, which has solution $y = c_1 \sinh(Kt + c_2)$. (The hyperbolic sine function is called `Sinh` in *Mathematica*.) Plot a numerical solution of Airy's equation together with a facsimile solution (with $K = 4$) on the interval $(14, 18)$. In analogy with part (b), you have to choose values for c_1 and c_2 in the facsimile solution.

   (d) Plot the numerical solution of Airy's equation on the interval $(-20, 2)$. What does the graph suggest about the frequency and amplitude of oscillations as $t \to -\infty$? Could any of that information have been predicted from the facsimile analysis?

2. Consider Bessel's equation of order zero

$$t^2 y'' + ty' + t^2 y = 0 \tag{D.1}$$

with initial conditions $y(0) = 1$, $y'(0) = 0$. The solution is the Bessel function of order zero of the first kind, $J_0(t)$. In this problem we solve equation (D.1) and two approximations of it with **DSolve** to learn about $J_0(t)$. Strictly speaking, this equation has a singularity at $t = 0$. However, this is one instance of a solution to a linear equation that exists outside the expected domain of definition. The singularity causes no difficulty in this problem.

(a) For t close to 0, t^2 is very small compared to t. The equation (D.1) is therefore approximately $ty' = 0$. Solve this equation with the preceding initial conditions. What does this "facsimile" solution to the original problem suggest to you about the behavior of J_0 near near $t = 0$?

(b) For t large and positive, t is small compared to t^2; and so we may approximate (D.1) by the equation $t^2(y'' + y) = 0$. Solve this equation with the same initial conditions. What does this suggest about the nature of the function J_0 for large t?

(c) Still thinking of t as large and positive, rewrite (D.1) in the form

$$y'' + \frac{1}{t}y' + y = 0.$$

If $t \approx K \gg 0$, we might approximate the equation by the constant coefficient equation

$$y'' + \frac{1}{K}y' + y = 0.$$

Choose a specific value for K, say $K = 100$, and solve this equation and see what further information you obtain about $J_0(t)$ for large positive t.

(d) Solve Bessel's equation with the given initial conditions, and plot the solution. Are the conclusions of your analysis confirmed?

3. This and some of the following problems concern models for the motion of a pendulum, which consists of a weight attached to a rigid arm of length L that is free to pivot in a complete circle. Neglecting friction and air resistance, the angle $\theta(t)$ that the arm makes with the vertical direction satisfies the differential equation

$$\theta''(t) + \frac{g}{L} \sin(\theta(t)) = 0, \tag{D.2}$$

where $g = 32.2$ ft/sec^2 is the gravitational acceleration constant. We will assume the arm has length 32.2 ft and so replace (D.2) by the simpler form

$$\theta'' + \sin \theta = 0. \tag{D.3}$$

(Alternatively, one can rescale time, replacing t by $\sqrt{g/L}\,t$, to convert (D.2) to (D.3).) For motions with small displacements (θ small), $\sin\theta \approx \theta$, and (D.3) can be approximated by the linear equation

$$\theta'' + \theta = 0. \tag{D.4}$$

This equation has general solution $\theta(t) = A\cos(t - \delta)$, with *amplitude A* and *phase shift δ*. Hence all the solutions to the linear approximation (D.4) have period 2π, independent of the amplitude A. In this problem we consider solutions of equation (D.3) satisfying the initial conditions $\theta(0) = A$, $\theta'(0) = 0$. If $|A| < \pi$, these solutions are periodic. However, in contract to the linear equation (D.4), their periods depend on the amplitude A. We do expect that, for small displacements A, the solutions to the pendulum (D.3) will have periods close to 2π.

(a) Investigate how the period depends on the amplitude A by plotting a numerical solution of equation (D.3) using initial conditions $\theta(0) = A$, $\theta'(0) = 0$ on an appropriate interval for various A. Estimate the periods of the pendulum for the amplitudes $A = 0.1, 0.7, 1.5$, and 3.0. Confirm these results by displaying the displacements at a sequence of times, and finding the time at which the pendulum returns to its original position.

(b) The period is given by the formula

$$T = 4 \int_0^{\pi/2} \frac{d\phi}{\sqrt{1 - k^2 \sin^2 \phi}},$$

where $k = \sin(A/2)$. This formula might be derived in your text; it can be found in Section 9.3 of Boyce and DiPrima, Problem 29. The integral is called an *elliptic integral*. It cannot be evaluated by an elementary formula, but it can be evaluated numerically using *Mathematica*. Calculate the period for the values of A we are considering. Do the values agree with those obtained in part (a)?

(c) Redo the numerical calculations in part (a) with different tolerances, choosing the tolerances so that the values you get agree with those calculated in part (b).

(d) How does the period depend on the amplitude of the initial displacement? For A small, is the period close to 2π? What is happening to the accuracy of the linear approximation as the initial displacement increases?

4. In this problem, we'll look at what the pendulum does for various initial velocities (*cf.* Problem 3).

(a) Numerically solve the differential equation (D.3) using initial conditions $\theta(0) = 0$, $\theta'(0) = 1$. Solve equation (D.4) with the same initial conditions. Plot on the same graph the solutions to both the nonlinear equation (D.3) and the linear equation (D.4) on the interval from $t = 0$ to $t = 40$, and compare the two. Be clear about which curve is the nonlinear solution and which is the linear solution.

(b) Repeat part (a) and compare the linear and nonlinear solutions for each of the following values of the initial velocity v: 1, 1.99, 2, 2.01. For the (numerical) nonlinear solution, interpret what the graph indicates the pendulum is doing physically. What do you think the exact solution does in each case?

5. In this problem, we'll investigate the effect of damping on the pendulum, using the model

$$\theta'' + b\theta' + \sin\theta = 0.$$

- Define a *Mathematica* function of b that will find a numerical solution of this differential equation with the initial conditions $\theta(0) = 0$, $\theta'(0) = 4$.
- For $b = 0.5$, 1, and 2, plot these solutions from $t = 0$ to $t = 20$.
- Do the same for the linear approximation

$$\theta'' + b\theta' + \theta = 0.$$

Compare the linear and nonlinear behavior for these values of b. Interpret what is happening *physically* in each case; *i.e.*, describe explicitly what the graph says the pendulum is doing.

6. In this problem, we'll look at the effect of a periodic external force on the pendulum, using the model

$$\theta'' + 0.05\theta' + \sin\theta = 0.3\cos\omega t \qquad\qquad (D.5)$$

(*cf.* Problems 3–5). We have chosen a value for the damping coefficient that is more typical of air resistance than the values in the previous problem.

- Write a *Mathematica* function of ω that numerically solves this differential equation with the initial conditions $\theta(0) = 0$, $\theta'(0) = 0$.
- For the following values of the frequency $\omega = 0.6, 0.8, 1.0, 1.2$ plot the solutions from $t = 0$ to $t = 60$.
- Do the same for the linear approximation

$$\theta'' + 0.05\theta' + \theta = 0.3\cos\omega t.$$

Compare the nonlinear and linear models for these values of the frequency ω. Which frequency moves the pendulum farthest away from its equilibrium position? For which frequencies do the linear and nonlinear equations have widely different behaviors? Which forcing frequency seems to induce resonance-type behavior in the pendulum? Graph that solution on a longer interval and decide whether the amplitude goes to infinity.

7. In this problem, we study the effects of air resistance.

(a) A paratrooper steps out of an airplane at a height of 1000 ft and after 5 seconds opens her parachute. Her weight, with equipment, is 195 lbs. Let $y(t)$ denote her height above the ground after t seconds. Assume that the force due to air resistance is $0.005y'(t)^2$ lbs in free fall and $0.6y'(t)^2$ lbs with the chute open. At what height does the chute open? How long does it take to reach the ground? At what velocity does she hit the ground? (This model assumes that air resistance is proportional to the *square* of the velocity and that the parachute opens instantaneously.)

(*Hint*: Pay attention to units. Recall that the mass of the paratrooper is $195/32$, measured in lb sec^2/ft. Here, 32 is the acceleration due to gravity, measured in ft/sec^2.)

(b) Let $v = y'$ be the velocity during the second phase of the fall (while the chute is open). One can view the equation of motion as an autonomous first order ODE in the velocity:

$$v' = -32 + \frac{192}{1950}v^2.$$

Make a qualitative analysis of this first order equation, finding in particular the critical or equilibrium velocity. This velocity is called the terminal velocity. How does the terminal velocity compare with the velocity at the time the chute opens and with the velocity at impact?

(c) Assume the paratrooper is safe if she strikes the ground at a velocity within 5% of the terminal velocity in (b). Except for the initial height, use the parameters in (a). What is the lowest height from which she may parachute safely? (*Please do not try this at home!*)

8. In many applications, second order equations come with *boundary conditions* rather than initial conditions; see Section 9.2. For example, consider a cable that is attached at each end to a post, with both ends at the same height (let us call this height $y = 0$). If the posts are located at $x = 0$ and $x = 1$, the height y of the cable as a function of x satisfies the differential equation

$$y'' = c\sqrt{1 + (y')^2},$$

with boundary conditions $y(0) = 0$, $y(1) = 0$. The constant c depends on the length of the cable; for this problem we'll use $c = 1$.

(a) Try solving the boundary value problem with **DSolve**; see the discussion in Section 9.2.

(b) Solve this problem numerically with the shooting method. Plot the solution on $[0, 1]$, and determine the maximum dip in the cable.

(c) Solve the auxiliary initial value problem using **DSolve**, and then find the critical value for the parameter s using **FindRoot**.

(d) Solve this boundary value problem using **NDSolve**.

Show that the solutions obtained in parts (a), (b), (c), and (d) agree with each other.

9. As explained in the last problem, sometimes one is faced with a second order differential equation with boundary conditions. Existence and uniqueness of solutions for
   such problems are more complicated than for initial value problems, as you will see
   in this problem. Consider the simple boundary value problem

   $$y'' + \alpha^2 y = 0, \qquad y(0) = 0, \quad y(1) = 1, \tag{D.6}$$

   where $\alpha > 0$ is a parameter.

   (a) Solve the problem exactly with **DSolve** using only the left-hand condition
       $y(0) = 0$. Since this is a second order equation and you have not specified
       $y'(0)$, the answer should involve an undetermined constant C[2]. Solve for
       this constant (in terms of α) using **Solve** in order to satisfy the right-hand
       condition $y(1) = 1$. For what values of α does a solution exist?

   (b) Redo part (a), but now modifying the right-hand condition in (D.6) to $y(1) = 0$.
       This time there is always a solution (namely $y = 0$), since the equation is now
       linear and homogeneous, but sometimes it is non-unique. For what values of α
       do you get more than one solution of the BVP? How much non-uniqueness is
       there in the solution?

   (c) What happens to the solution of (D.6) (as computed in part (a)) when $\alpha \to \pi$?
       What do you observe about the solution as computed with
       NDSolve for α close to π, say 3.1415926? What about $\alpha = 3.14159265359$?
       How do the plots of the solutions in the two cases differ, and how can you
       explain this?

10. The problem of finding the function $u(x)$ satisfying the boundary value problem

    $$\begin{cases} a(x)u''(x) + a'(x)u'(x) = f(x), & 0 \le x \le 1 \\ u(0) = u(1) = 0, \end{cases} \tag{D.7}$$

    arises in studying the longitudinal displacements in a longitudinally loaded elastic
    bar. The bar is of length 1, its left end is at $x = 0$, and its right end at $x = 1$. In
    the differential equation above, $f(x)$ represents the external force on the bar (which
    is assumed to be longitudinal, *i.e.*, directed along the bar), $a(x)$ represents both the
    elastic properties and the cross-sectional area of the bar, and $u(x)$ is the longitudinal
    displacement of the bar at the point x. The function $a(x)$ may be constant; this is
    the case if neither the elastic properties nor the cross-sectional area depends on the
    position x in the rod, *i.e.*, if the rod is uniform. But if the rod is not uniform, then
    $a(x)$ is not constant, and the equation has variable coefficients. Similarly, $f(x)$ will
    be constant only if the external force is applied uniformly along the rod.

    (a) Take $a(x) = 1 + x$ and $f(x) = 5\sin^2(2\pi x)$. Here the force is applied sym-
        metrically and is strongest at $x = 0.25$ and $x = 0.75$, as if someone is hanging
        by both hands from a relatively short rod. Use **DSolve** to find a solution $u(x)$
        to the boundary problem (D.7), and plot it on the interval $[0, 1]$. Based on the

result, do you think the rod is more flexible where $a(x)$ is larger or where it is smaller?

(b) Now suppose $a(x) = 1 + \exp(x)$ and $f(x)$ is the same as in part (a). Find and plot the corresponding solution $u(x)$ to (D.7). Since *Mathematica* cannot solve this equation symbolically, we will use a numerical method. Use the shooting method with **NDSolve** to find the value of $u'(0)$ that leads to a solution satisfying the condition $u(1) = 0$. By trial and error, find $u'(0)$ to at least two decimal places, and graph the resulting solution on $[0, 1]$. What is the maximum displacement, and where does it occur?

(c) Repeat part (b) using **NDSolve** directly on the boundary value problem, instead of the shooting method; see discussion in Section 9.2.

11. This problem is based on Problem 32 in Boyce and DiPrima, Section 3.7. Consider a frictionless mass-spring system as in Figure 3.7.10 in Boyce and DiPrima (standard mass attached to a spring on a frictionless table). Suppose the restoring force of the spring is not given by Hooke's Law, but instead is of the form

$$F = -(ky + \epsilon y^3).$$

If $\epsilon = 0$, the assumption amounts to Hooke's Law, but in this problem we shall focus on $\epsilon \neq 0$ (either positive or negative). If we have air resistance present with damping coefficient γ, then the equation of motion (assuming the displacement is indicated by the variable y) becomes

$$my'' + \gamma y' + ky + \epsilon y^3 = 0.$$

Henceforth we shall normalize by assuming that $m = k = 1$ and $\gamma = 0$, and then take the initial conditions

$$y(0) = 0, \quad y'(0) = 1.$$

(a) Plot the solution when $\epsilon = 0$. What is the amplitude and period of the solution?

(b) Let $\epsilon = 0.1$. Plot a numerical solution. Is the motion periodic? Estimate the amplitude and period.

(c) Repeat part (b) for $\epsilon = 0.2$, and then for $\epsilon = 0.3$.

(d) Plot your estimated values of the amplitude A and period T as functions of ϵ. How do A and T depend on ϵ?

(e) Repeat parts (b)–(d) for negative values of ϵ.

12. In this problem we study how solutions of the initial value problem

$$\frac{d^2y}{dt^2} + 0.15\frac{dy}{dt} - y + y^3 = 0, \quad y(0) = c, \quad y'(0) = 0$$

depend on the initial value c.

(a) Plot a numerical solution of this equation from $t = 0$ to $t = 40$ for each of the initial values $c = 0.5, 1, 1.5, 2, 2.5$. Describe how increasing the initial value of y affects the solutions, both in terms of their limiting behavior and their general appearance.

(b) Plot all five solutions on one graph. Would such a graph be possible for solutions of a first order differential equation? Why or why not?

13. In this problem, we consider the long-term behavior of solutions of the initial value problem

$$\frac{d^2y}{dt^2} + 0.2\frac{dy}{dt} - y + y^3 = 0.3\cos(\omega t), \quad y(0) = 0, \quad y'(0) = 0 \qquad \text{(D.8)}$$

for various frequencies ω in the forcing term.

Plot a numerical solution of (D.8) from $t = 0$ to $t = 100$ for each of the eight frequencies $\omega = 0.8, 0.9, \ldots, 1.4, 1.5$.

Describe and compare the different long-term behaviors you see. Due to the forcing term, all solutions will oscillate, but pay particular attention to the magnitude of the oscillations and to whether or not there is a periodic pattern to them. Are there any similarities between your results for this nonlinear system and the phenomenon of resonance for linear systems with periodic forcing?

14. In this problem, we study the zeros of solutions of the second order differential equation

$$y'' + (3 - \cos t)y = 0. \qquad \text{(D.9)}$$

(a) Compute and plot several solutions of this equation with different initial conditions: $y(0) = c$, $y'(0) = d$. To be specific, choose three different values for the pair c, d and plot the corresponding solutions over $[0, 20]$. By inspecting your plots, find a number L that is an upper bound for the distance between successive zeros of the solutions. Then find a number l that is a lower bound for the distance between successive zeros.

(b) Information on the zeros of solutions of linear second order ODEs can be obtained from the Sturm Comparison Theorem (see Theorem 9.1 in Chapter 9). By comparing (D.9) with the equation $y'' + 2y = 0$, you should be able to get a value for L, and by comparing (D.9) with $y'' + 4y = 0$, you should be able to find l. How do these values compare with the values obtained in part (a)? (You will find it useful to note that the general solution of $y'' + ky = 0$, where k is a positive constant, can be written as $y = R\cos(\sqrt{k}x - \delta)$, with $R \geq 0$, δ arbitrary. R is called the *amplitude* and δ the *phase shift*.)

(c) Plot a solution of (D.9) and a solution of $y'' + 2y = 0$ on the same graph, and verify that between any two zeros of the latter solution, there is at least one zero of the solution of (D.9).

15. This problem is based on Problem 18 in Boyce and DiPrima, Section 3.8. Consider the initial value problem

$$u'' + u = 3\cos(\omega t), \quad u(0) = 0, \quad u'(0) = 0.$$

(a) Find the solution (using **DSolve**). For $\omega = 0.5, 0.6, 0.7, 0.8, 0.9$, plot the solution curves on the interval $0 \le t \le 15$. Note that $\omega_0 = 1$ is the natural frequency of the homogeneous equation. Describe how the solution curves change as ω gets closer to 1.

(b) Note that the formula you found in part (a) is invalid when $\omega = 1$. Find and plot the solution curve for $\omega = 1$ on the interval $0 \le t \le 15$. Based on the discussion of forced vibrations in your text, what phenomenon should be exhibited for this value of ω? Corroborate your answer by plotting on a longer interval.

(c) Plot the solution for $\omega = 0.9$ on a longer interval and compare it with the solution from part (b). What phenomenon is exhibited by the curve for $\omega = 0.9$?

16. A solution of a second order linear differential equation is called *oscillatory* if it changes sign infinitely many times and *nonoscillatory* if it changes sign only finitely many times. In this problem, we will be interested in determining the oscillatory nature of nonzero solutions to some second order linear ODEs. First consider the equation $y'' + ky = 0$, where k is a constant. If $k > 0$, it has the general solution $y = R\cos(\sqrt{k}x - \delta)$, with $R \ge 0, \delta$ arbitrary. The constant R is called the *amplitude*, and δ is called the *phase shift*. From this formula we see that every solution $y(x)$ has infinitely many zeros, and hence changes sign infinitely many times, *i.e.*, is oscillatory. If $k < 0$, the general solution is $y(x) = c_1 e^{\sqrt{-k}x} + c_2 e^{-\sqrt{-k}x}$, while if $k = 0$, the general solution is $y(x) = c_1 x + c_2$. These solutions change sign at most once, and so are nonoscillatory.

Next, consider Airy's equation

$$y'' = xy, \tag{D.10}$$

which arises in various applications. Since (D.10) has a variable coefficient, we cannot study it by elementary methods; in particular, we cannot find the general solution in terms of elementary functions as we did for the constant coefficient equation. We will instead first make a graphical study of the solutions of (D.10) and then study the solutions using the Sturm Comparison Theorem (see Chapter 9).

(a) Compute and plot several solutions of (D.10) with different initial conditions $y(0) = c$, $y'(0) = d$ over the interval $[-10, 5]$. Use different combinations of positive, negative, and zero values for c and d. What can you say in general about the zeros of the solution? Where do they occur on the x-axis, and how far apart are the successive zeros of a solution? Are the solutions oscillatory as $x \to \infty$? As $x \to -\infty$?

(b) Information on the zeros of solutions of linear second order ODEs, and hence information on their oscillatory nature, can also be obtained from the Sturm Comparison Theorem. By comparing (D.10) with the equation $y'' + by = 0$ for $x \leq -b < 0$, what do you learn about zeros on the negative x-axis and their spacing, and hence about the oscillatory nature of solutions of (D.10) for $x \leq 0$? By comparing (D.10) with $y'' = 0$ for $x > 0$, what do you learn about the oscillatory nature of solutions for $x > 0$? How many zeros could a solution have on the positive x-axis? Do the graphs you plotted in part (a) agree with these results?

17. In this problem, we study solutions of the parabolic cylinder equation

$$y'' + \left(n + \frac{1}{2} - \frac{x^2}{4}\right) y = 0, \tag{D.11}$$

which arises in the study of quantum-mechanical vibrations. Since the equation (D.11) is unchanged if x is replaced by $-x$, any solution function will be symmetric with respect to the y-axis. Therefore, we focus our attention on $x \geq 0$.

(a) Find the corresponding first order equation for $z = \arctan(y/y')$, as described in Section 9.4.

(b) For $n = 1$, plot the direction field for the z equation from $x = 0$ to $x = 10$. (Remember to use $-\pi/2 \leq z \leq \pi/2$.) Based on the plot, predict what the solutions y to the parabolic cylinder equation look like near $x = 0$ and for larger x. Is there a value of x around which you expect their behavior to change?

(c) Now solve the parabolic cylinder equation numerically for $n = 1$ with the two sets of initial conditions $y(0) = 1$, $y'(0) = 0$ and $y(0) = 0$, $y'(0) = 1$. Plot the two solutions on the same graph. (It will probably help to change the range on the plot.) Do the solutions behave as you expected?

(d) Repeat parts (b) and (c) for $n = 5$. Point out any differences from the case $n = 1$.

(e) Repeat parts (b) and (c) for $n = 15$. Discuss how the solutions are changing as n increases.

(f) Now consider the three solutions (for $n = 1$, 5, and 15) with $y(0) = 0$. Point out similarities and differences. By drawing an analogy with Airy's equation, argue from the direction fields that, for any n, exactly one solution function decays for large x while all others grow. Do you have enough graphical evidence from the numerical plots to conclude that the solution corresponding to the initial data $y(0) = 0$, $y'(0) = 1$ is that function? Why or why not? (*Hint*: Look at previous discussions of stability of solutions as a guide.)

18. Consider Bessel's equation of order zero

$$x^2 y'' + xy' + x^2 y = 0.$$

(a) Compute and plot several solutions of this equation with different initial conditions: $y(0.1) = c, y'(0.1) = d$. To be specific, choose three different values for the pair c, d and plot the corresponding solutions over $[0.1, 20]$. (Why isn't it a good idea to use $x = 0$ in the initial conditions?) By inspecting your plots, find a number L that is an upper bound for the distance between successive zeros of the solutions. Then find a number l that is a lower bound for the distance between successive zeros.

(b) Now confirm your findings with the Sturm Comparison Theorem. It doesn't apply directly, but if we introduce the new function $z(x) = x^{1/2}y(x)$, Bessel's equation becomes

$$z'' + (1 + 1/(4x^2))z = 0,$$

which has the form of the equation in the Comparison Theorem. Since y will have a zero wherever z has a zero, we can study the zeros of y by studying the zeros of z. By comparing the equation for z with the equation $z'' + z = 0$, determine an upper bound L on the distance between successive zeros of any solution of Bessel's equation.

(c) Let a be a positive number. For $x \in [a, \infty)$, the quantity $1 + 1/(4x^2)$ is less than or equal to the constant $1 + 1/(4a^2)$. By making an appropriate comparison, determine a lower bound l on the distance between successive zeros of any solution of Bessel's equation (for $x > a$). What is the limiting value of l as a goes to ∞? Approximately how far apart are the zeros when x is large? Did your graphical study lead to comparable values for l and L?

19. In this problem, we study solutions of Bessel's equation of order n,

$$y'' + \frac{1}{x}y' + \left(1 - \frac{n^2}{x^2}\right)y = 0, \tag{D.12}$$

for $n > 0$. Solutions of this equation, called *Bessel functions* of order n, are used in the study of vibrations and waves with circular symmetry. Since (D.12) is unchanged if x is replaced by $-x$, we focus our attention on $x \geq 0$.

(a) Find the corresponding first order equation for $z = \arctan(y/y')$, as described in Section 9.4.

(b) For $n = 1$, plot the direction field for the z equation from $x = 0$ to $x = 20$. (Remember to use $-\pi/2 \leq z \leq \pi/2$.) Based on the plot, predict what the Bessel functions of order 1 look like for small x and then for large x. Is there a value of x around which you expect their behavior to change?

(c) Now plot the Bessel functions `BesselJ[n, x]` and `BesselY[n, x]` for $n = 1$ on the same graph. (It will probably help to change the range on the plot.) Do the solutions behave as you expected?

(d) Repeat parts (b) and (c) for $n = 5$. Point out any differences from the case $n = 1$.

(e) Repeat parts (b) and (c) for $n = 15$. Discuss how the solutions are changing as n increases.

20. Consider the mass-spring system with dry friction depicted in Figure D.1. This sys-

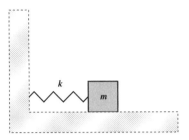

Figure D.1: Mass-Spring System with Friction

tem is governed by the following equations:

$$\begin{cases} my'' = -ky - \mu mg \operatorname{sign}(y'), & \text{unless } y' = 0 \text{ and } |ky| < \mu mg, \\ y'(t) \equiv 0, \quad t \geq t_0 & \text{if } y'(t_0) = 0 \text{ and } |ky(t_0)| < \mu mg. \end{cases} \quad \text{(D.13)}$$

The first equation in (D.13) governs *kinetic friction*, and says that the object of mass m is subject to a restoring force opposite and proportional to the displacement y, with proportionality constant the spring constant k, as well as to a friction force opposite to the direction of motion, given by the friction coefficient μ times the force of gravity on the mass. The second equation in (D.13) governs *static friction*, and says that if the mass is instantaneously not moving at time t_0 and the force exerted on it by the spring is not big enough, then it will never move at all. In this model we are assuming for simplicity (even though this is usually not the case) that the coefficients of static and kinetic friction are equal. (In a more realistic model, the μ in the second equation would be larger than the μ in the first equation.) We are also assuming that the force of friction is independent of velocity in the kinetic friction case, which is not 100% correct, but is a reasonable first approximation to the truth. Note that the equations (D.13) really only depend on two constants, $d = \mu g$ and $\omega = \sqrt{k/m}$. When $\mu = 0$ (case of no friction), they reduce to the usual spring equation $y'' + \omega^2 y = 0$ with angular frequency ω. On the other hand, in the limiting case where $\mu \to \infty$, the object comes to an immediate stop and then doesn't move at all.

(a) The first equation in (D.13) can be rewritten as

$$y'' + \omega^2 y + d \operatorname{sign}(y') = 0.$$

Suppose we denote the initial data by $y_0 = y(0)$, $v_0 = y'(0)$. Solve the IVP numerically in *Mathematica* when $y_0 = 4$, $v_0 = 0$, $\omega = 1$, and $d = 0.75$. (The sign function is implemented in *Mathematica* as `Sign`.) Draw two graphs: on one, graph the displacement y for $0 \leq t \leq 10$; on the second, draw a phase diagram for y, y' over the same time interval. Now note that the second equation in (D.13) says that the mass comes to rest as soon as it reaches a point in the (y, y') phase plane on the horizontal axis (so $y' = 0$) and with

$$|y| < \frac{\mu mg}{k} = \frac{d}{\omega^2}$$

Looking at your phase plot, where does the mass appear to come to rest?

(b) Now let's try to get a more precise value for where the mass comes to rest, using event detection for the numerical ODE solver, as discussed in Section 8.12.1. Your phase plot in (a) should show that the trajectory in the phase plane crosses the horizontal axis several times. The first crossing that satisfies the condition $|y| < d/\omega^2$ is after $t = 9$. You can get a more precise value for the final value of y as follows. First compute y and y' at $t = 9$. Then use these as new initial conditions for another call of `NDSolve`, but this time with `Method` $\rightarrow$ `EventLocator` set to detect the first place (after $t = 9$) where $y' = 0$.

(c) Repeat the calculations of parts (a) and (b) but with $y_0 = 3$, then with $y_0 = 2$.

(d) Now change the initial data to $y_0 = 0$, $v_0 = 5$ and leave $\omega = 1, d = 0.75$. What kind of motion ensues? (*Hint*: You may need to lengthen the time interval.)

(e) Find (to the nearest tenth) the value of v_0 (with $y_0 = 0$, $\omega = 1, d = 0.75$) such that initial velocities below or equal to that value do not propel the mass out of the "friction well" $-0.75 \leq y \leq 0.75$, but initial velocities above it propel it so that it goes beyond the well once, but when it returns to the well, it does not escape it again.

Chapter 10

Series Solutions

A primary theme of this book is that numerical, geometric, and qualitative methods can be used to study solutions of differential equations, even when we cannot find an exact formula solution. Of course, formula solutions are extremely valuable, and there are many techniques for finding them. For instance, techniques for finding formula solutions of second order linear differential equations include:

- The exponential substitution, which leads to solutions of an arbitrary constant coefficient homogeneous equation.

- The method of *reduction of order*, which produces a second linearly independent solution of a homogeneous equation when one solution is already known.

- The method of *undetermined coefficients*, which solves special kinds of inhomogeneous equations with constant coefficients.

- The method of *variation of parameters*, which yields the solution of a general inhomogeneous equation, given a fundamental set of solutions to the homogeneous equation.

Each of these techniques reduces the search for a formula solution to a simpler problem in algebra or calculus: finding the root of a polynomial, computing an antiderivative, or solving a pair of simultaneous linear equations.

The process of finding formulas for exact solutions of equations, either by hand or by computer, is called *symbolic computation*. The **DSolve** command incorporates all of the techniques listed above. It enables us to find exact solutions more rapidly and more reliably than we could by hand.

There are many differential equations that do not yield to the techniques listed above. By using more advanced ideas from calculus, however, we can find exact or approximate solutions for a wider class of differential equations. In this chapter and the next, we discuss two such calculus-based techniques for finding solutions of differential equations: *series solutions* and *Laplace transforms*. The method of series solutions constructs power series solutions of linear differential equations with variable coefficients; in many examples, we compute only finitely many terms of the power series and therefore obtain Taylor polynomial approximations of the solutions. The theory of Laplace transforms (discussed in the next chapter) enables us to solve constant coefficient linear differential equations with discontinuous inhomogeneous terms. These are sophisticated techniques, involving improper integrals, complex variables, and power series.

For each technique, we describe the use of *Mathematica* to automate the process of computing a symbolic solution to a differential equation or to an associated initial value problem. Unlike our previous uses of `DSolve` to find symbolic solutions, the methods discussed in these chapters involve a *sequence* of *Mathematica* commands. As with numerical and graphical methods, computationally intensive symbolic methods become more tractable when done with the aid of a mathematical software system.

10.1   Series Solutions

Consider the second order homogeneous linear differential equation with variable coefficients

$$a(x)y''(x) + b(x)y'(x) + c(x)y(x) = 0. \tag{10.1}$$

Suppose the coefficient functions a, b, and c are *analytic* (*i.e.*, have convergent power series representations) in a neighborhood of a point $x = x_0$. For simplicity, we shall assume that $x_0 = 0$, but everything we say remains valid for any point of analyticity. We begin by dividing by $a(x)$ to normalize the equation. This causes no problems as long as $a(0) \neq 0$. In this case, we refer to the origin as an *ordinary point* for the equation. If we set $p(x) = b(x)/a(x)$ and $q(x) = c(x)/a(x)$, then equation (10.1) can be rewritten in the form

$$y''(x) + p(x)y'(x) + q(x)y(x) = 0. \tag{10.2}$$

Equations (10.1) and (10.2) have the same solutions. We will search for a power series solution of the form

$$y(x) = \sum_{n=0}^{\infty} a_n x^n. \tag{10.3}$$

If we can find such a solution, then $y(0) = a_0$ and $y'(0) = a_1$. Thus, if we are given initial values, we can find the first two coefficients of the power series solution.

By substituting the infinite series (10.3) into equation (10.2), expanding p and q as power series, and combining terms of the same degree, we obtain a *recursion relation* for the coefficients a_n. In other words, for each $n \geq 2$, we obtain an algebraic equation for a_n involving $a_0, a_1, \ldots, a_{n-1}$ and the coefficients of the power series of p and q. Sometimes, we can solve the recursion relation in *closed form* by finding an algebraic formula for a_n. This formula would involve known functions of n (such as $n!$ or powers of n) and a_0 and a_1, but no other coefficients of lower degree. If no initial data are given, then a solution to the recursion relation produces a general solution of the differential equation with arbitrary constants a_0 and a_1.

Solving a recursion relation in closed form is analogous to finding an exact formula solution for a differential equation. In many cases, we cannot solve the recursion relation.

Nevertheless, we can still use the recursion relation to compute as many coefficients as we wish. As long as we stay close to the origin, the leading terms should give a good approximation to the full power series solution. One must be careful, however, because the power series solution will be valid only inside its radius of convergence. The radius of convergence will be at least as large as the distance from the origin to the nearest singularity of p or q. When the recursion relation is solvable in closed form, we can use a standard test from calculus to compute the radius of convergence precisely.

We can use *Mathematica* to automate the process of finding power series solutions. In fact, *Mathematica* can solve some recursion relations in closed form. You can explore that possibility by experimenting with **RSolve**. Here, however, we shall describe a straight-forward technique for computing the leading coefficients explicitly.

Example 10.1 We would like to compute the first five terms in the power series approximation of the initial value problem

$$y'' - xy' - y = 0, \qquad y(0) = 2, \qquad y'(0) = 1. \tag{10.4}$$

The **Series** command generates a power series for an analytic function in a neighborhood of a point to any number of terms. For example, the command

```
Series[Exp[x],{x, 0, 4}]
```

gives the first five terms (*i.e.*, terms up to order four) of the Maclaurin series for e^x, namely

$$1 + x + \frac{x^2}{2} + \frac{x^3}{6} + \frac{x^4}{24} + O[x]^5$$

The symbol $O[x]^5$ means terms of degree at least 5. We can also produce a formal series. For example,

```
formalseries = Series[y[x],{x, 0, 4}]
```

yields

$$y[0] + y'[0] x + \frac{1}{2}y''[0] x^2 + \frac{1}{6}y^{(3)}[0] x^3 + \frac{1}{24}y^{(4)}[0] x^4 + O[x]^5$$

We can even apply the **Series** command to the left-hand side of a differential equation, as follows:

```
odeseries1 = Series[y''[x] - x y'[x] - y[x],{x, 0, 4}]
```

The result is

$$(-y[0] + y''[0]) + (-2y'[0] + y^{(3)}[0]) x + \frac{1}{2}(-3y''[0] + y^{(4)}[0]) x^2 +$$

$$\frac{1}{6}(-4y^{(3)}[0] + y^{(5)}[0]) x^3 + \frac{1}{24}(-5y^{(4)}[0] + y^{(6)}[0]) x^4 + O[x]^5$$

To produce a series solution to a differential equation, we first set the series expansion of the left-hand side of the differential equation equal to zero, incorporate the initial conditions, and apply the **Solve** command. The output will be a list of replacement rules containing the coefficients of the series solution.

```
coeffs1 = Solve[{odeseries1 == 0, y[0] == 2, y'[0] == 1}]
```

$$\{\{y[0] \to 2, y'[0] \to 1, y''[0] \to 2, y^{(3)}[0] \to 2, y^{(4)}[0] \to 6,$$
$$y^{(5)}[0] \to 8, y^{(6)}[0] \to 30\}\}$$

Now we use these replacement rules to substitute for the coefficients in a formal power series:

```
seriessol1 = Series[y[x], {x, 0, 4}] /. First[coeffs1]
```

The result is

$$2 + x + x^2 + \frac{x^3}{3} + \frac{x^4}{4} + O[x]^5.$$

We can use **Normal** to convert the series solution to a polynomial by removing the $O[x]^5$.

```
answer1 = Normal[seriessol1]
```

This produces the Taylor polynomial approximation of degree 4 to the solution of the initial value problem:

$$2 + x + x^2 + \frac{x^3}{3} + \frac{x^4}{4}.$$

The process we followed in this example can be automated by defining a *Mathematica* procedure **SeriesSol** that will work quite generally. Following the example above, we define:

```
SeriesSol[xvar_, yvar_, ode_, inits_, order_] :=
    Module[{odeseries, coeffs},
        odeseries = Series[ode, {xvar, 0, order}];
        coeffs = Solve[Join[{odeseries == 0}, inits]];
        Series[yvar[xvar], {xvar, 0, order}] /. First[coeffs]]
```

Here the parameters **xvar** and **yvar** are the independent and dependent variables, respectively; **ode** is the differential equation (or rather what is set equal to 0); **inits** is the list of initial conditions (if there are any); and **order** is the desired degree of approximation. The command **Module** allows us to define *local* variables inside the procedure—in this case, **odeseries** and **coeffs**. We solve for **coeffs** using the union of the equation **odeseries == 0** and the initial conditions, then substitute back into the formal series expansion of **yvar[xvar]**.

We can test this procedure on the example above:

```
SeriesSol[x, y, y''[x] - x*y'[x] - y[x],
    {y[0] == 2, y'[0] == 1}, 5]
```

$$2 + x + x^2 + \frac{x^3}{3} + \frac{x^4}{4} + \frac{x^5}{15} + O[x]^6.$$

While the **DSolve** command will find exact solutions to many homogeneous second order equations with polynomial coefficients, equations with coefficients that are not polynomials generally require series methods. When **DSolve** does find solutions, they often involve special functions like **BesselJ** and **BesselY**. So far, we have focused on linear homogeneous second order differential equations. The method of series solutions can also be used for inhomogeneous equations, higher order equations, and nonlinear equations. As we've noted before, most differential equations cannot be solved in terms of elementary functions. For many of these equations, a series solution is the only formula solution available.

Here's an example using a non-linear equation.

Example 10.2 We would like to compute the terms out to degree 7 in the power series approximation of the solution to the initial value problem

$$y' + xy^2 - y - 1 = 0, \qquad y(0) = 1. \tag{10.5}$$

(**DSolve** can solve this initial value problem, but the solution is quite complicated and involves all of the special functions **AiryAi**, **AiryBi**, **AiryAiPrime**, **AiryBiPrime**, and **Gamma**.) Using our procedure above, this is a snap:

```
SeriesSol[x, y, y'[x] + x*y[x]^2 - 1, {y[0] == 1}, 7]
```

$$1 + x - \frac{x^2}{2} + \frac{2x^3}{3} + \frac{7x^5}{15} + \frac{13x^6}{72} - \frac{8x^7}{35} + O[x]^8$$

10.2 Singular Points

In the discussion above, we considered equations of the form

$$a(x)y'' + b(x)y' + c(x)y = 0,$$

where $a(0) \neq 0$, and we transformed the equation into the form

$$y'' + p(x)y' + q(x)y = 0$$

by dividing by $a(x)$. If $a(0) = 0$, dividing by $a(x)$ may result in one or both of $p(x)$ and $q(x)$ being singular at $x = 0$. In this case, we say that the equation has a *singularity* at $x = 0$.

A prototype is the Euler equation:

$$x^2 y'' + bxy' + cy = 0,$$

where b and c are constants. Dividing by x^2 yields the equation

$$y'' + \frac{b}{x}y' + \frac{c}{x^2}y = 0, \tag{10.6}$$

so $p(x) = b/x$ and $q(x) = c/x^2$. Both $p(x)$ and $q(x)$ are singular at $x = 0$. Such isolated singularities, where the singularity of p is no worse than $1/x$ and the singularity of q is no worse than $1/x^2$, are called *regular singular points*. More precisely, we say that a function $f(x)$ has a *pole of order n* at x_0 if $\lim_{x \to x_0} f(x) = \infty$, if $(x - x_0)^n f(x)$ has a convergent power series expansion around $x = x_0$, and if n is the smallest integer such that $\lim_{x \to x_0}(x - x_0)^n f(x)$ is finite. A homogeneous linear differential equation $y^{(n)} + p_1(x)y^{(n-1)} + \cdots + p_n(x)y = 0$ is said to have a regular singular point at x_0 if it is singular at x_0, and p_k has a pole of order at most k at x_0.

Note that the Euler equation (10.6) has x^r as a solution, provided that r satisfies the *indicial equation*, $r(r - 1) + br + c = 0$. The solution procedure for general second order equations with regular singular points is based on this example as a prototype. We suppose, for simplicity, that the singular point is $x = 0$, and we look for solutions on the interval $(0, \infty)$. Let $p_0 = \lim_{x \to 0} xp(x)$ and $q_0 = \lim_{x \to 0} x^2 q(x)$. Let r_1 and r_2 be the roots of the indicial equation, $r(r - 1) + p_0 r + q_0 = 0$, with $r_1 \geq r_2$ if the roots are real. We look for a solution of the differential equation of the form

$$y_1(x) = x^{r_1} u(x),$$

where $u(0) = 1$. If $r_1 - r_2$ is not an integer, then we look for a second solution of the form

$$y_2(x) = x^{r_2} v(x),$$

where $v(0) = 1$. If $r_1 = r_2$, then we look for a second solution of the form

$$y_2(x) = y_1(x)\ln(x) + x^{r_1} v(x),$$

where $v(0) = 0$. If $r_1 - r_2$ is a positive integer, then we look for a second solution of the form

$$y_2(x) = ay_1(x)\ln(x) + x^{r_2} v(x),$$

where a is a constant (which might be 0) and $v(0) = 1$. The functions $u(x)$ and $v(x)$ will be analytic at 0 and therefore will have power series expansions at $x = 0$. These solutions are called *Frobenius series* solutions.

Example 10.3 Consider the differential equation:

$$2xy'' + y' + xy = 0. \tag{10.7}$$

The origin is a regular singular point, with $p_0 = 1/2$ and $q_0 = 0$. The indicial equation is $r(r - 1) + r/2 = 0$, which has roots $r_1 = 1/2$ and $r_2 = 0$. Hence, the differential equation will have one Frobenius series solution that is analytic, and one of the form

$$x^{1/2} \sum_{n=0}^{\infty} a_n x^n.$$

We will adapt the series solution method for ordinary points to calculate the leading terms of the Frobenius series corresponding to $r_1 = 1/2$. We will define $y(x)$ to be $\sqrt{x}\,u(x)$ and expand both y and the differential equation. Here is the sequence of commands:

```
y[x_] = √x*u[x];
    odeseries2 = Series[2 x*y''[x] +y'[x] +x*y[x], {x, 0, 5}];
    coeffs2 = Solve[{odeseries2 == 0, u[0] == 1}];
    seriessol2 = Series[y[x], {x, 0, 5}] /. First[coeffs2];
    answer2 = Normal[seriessol2]
```

The final result is

$$\sqrt{x} - \frac{x^{5/2}}{10} + \frac{x^{9/2}}{360} .$$

To obtain another linearly independent solution of the equation, we repeat this procedure using the other exponent, $r_2 = 0$.

```
y[x_] = v[x];
    coeffs3 = Solve[{odeseries2 == 0, v[0] == 1}];
    seriessol3 = Series[y[x], {x, 0, 5}] /. First[coeffs3];
    answer3 = Normal[seriessol3]
```

The final result is

$$1 - \frac{x^2}{6} + \frac{x^4}{168} .$$

This is the Taylor series approximation for a second linearly independent solution $y(x)$.

Now we can put our two solutions together. The general solution to the equation has the form

$$y(x) = C1\,\sqrt{x}\left(1 - \frac{1}{10}x^2 + \frac{1}{360}x^4 + O\left(x^6\right)\right)$$
$$+ C2\left(1 - \frac{1}{6}x^2 + \frac{1}{168}x^4 + O\left(x^6\right)\right).$$

Note that we did not specify an initial condition in this example, although we could have since the solutions have finite limits at 0. In general, specifying an initial condition (at the singular point) for a singular differential equation will cause **DSolve** to fail.

Exercise 10.1 Compute two more terms in each of the examples presented in this chapter.

It is important to note that for equations with regular singular points, the Frobenius series tells us how fast the solution blows up at the singularity; *e.g.*, the solution blows up like $1/x$, $x^{-1/2}$, *etc.* This information is not easily gleaned from a numerical solution. For equations with an irregular singular point, we cannot expect a Frobenius series solution to be valid, and other techniques (numerical or qualitative) must be used. Some of those techniques are addressed in Problem Set E.

Chapter 11

Laplace Transforms

A *transform* is a mathematical operation that changes a given function into a new function. Transforms are often used in mathematics to change a difficult problem into a more tractable one. In this chapter, we introduce the *Laplace transform*, which is especially useful for solving linear differential equations with constant coefficients and discontinuous inhomogeneous terms. The key feature of the Laplace transform is that (roughly speaking) it changes the operation of differentiation into the operation of multiplication. Thus the Laplace transform changes a differential equation into an algebraic equation. To solve a linear differential equation with constant coefficients, you apply the Laplace transform to change the differential equation into an algebraic equation, solve the algebraic equation, and then apply the inverse Laplace transform to transform the solution of the algebraic equation back into the solution of the differential equation.

The Laplace transform of a function f is a new function, denoted by F or by $\mathcal{L}(f)$, and defined as follows:

$$F(s) = \mathcal{L}(f)(s) = \int_0^\infty f(t)e^{-st}\,dt.$$

This transform is called an *integral transform* because it is obtained by integrating the function f against another function e^{-st}, called the *kernel* of the transform. The integral in question is an improper integral, so we have to make sure that it converges. Notice that whereas the argument s of the function F appears as a parameter in the integrand, the integration is with respect to the variable t. Also, the integral is over the domain $[0, \infty)$, so we assume that the function f is defined for $t \geq 0$.

To get a feel for the Laplace transform, we compute the Laplace transform of the function $g(t) = e^{at}$.

$$\begin{aligned}
\mathcal{L}(g)(s) &= \int_0^\infty e^{at}e^{-st}\,dt = \lim_{b\to\infty}\int_0^b e^{at}e^{-st}\,dt \\
&= \lim_{b\to\infty}\int_0^b e^{(a-s)t}\,dt = \lim_{b\to\infty}\left.\frac{e^{(a-s)t}}{(a-s)}\right|_0^b
\end{aligned}$$

$$= \lim_{b \to \infty} \left(\frac{e^{(a-s)b}}{(a-s)} - \frac{1}{a-s} \right)$$

$$= \begin{cases} 1/(s-a), & \text{if } s > a \\ +\infty, & \text{if } s \le a. \end{cases}$$

To avoid the infinite values, we say that the Laplace transform of e^{at} is defined only for $s > a$. A straightforward argument shows that if f is any piecewise continuous function on $[0, \infty)$ with the property that $|f(t)| \le Ke^{at}$, for some constant $K > 0$, then the improper integral defining the Laplace transform converges for $s > a$, and therefore $\mathcal{L}(f)(s)$ is defined at least for $s > a$. (We say that a function is *piecewise continuous* if it only has a discrete set of jump discontinuities.) If f satisfies an inequality of the form $|f(t)| \le Ke^{at}$, we say that f is of *exponential order*. Most functions one encounters in practice are of exponential order. In the rest of this chapter, we only consider piecewise continuous functions of exponential order. In particular, if f is a bounded function, then it satisfies the inequality $|f(t)| \le K = Ke^{0t}$ for some $K > 0$, so $\mathcal{L}(f)(s)$ is defined for all $s > 0$. More generally, any function whose growth is of exponential order has a Laplace transform that is defined for sufficiently large s.

Exercise 11.1 Compute the Laplace transforms of the functions $f(t) = 1$ and $\cos t$. (For $\cos t$, you must integrate by parts twice.)

We asserted that the Laplace transform changes differentiation into multiplication. This is a consequence of the integration by parts formula:

$$\begin{aligned} \mathcal{L}(f')(s) &= \int_0^\infty f'(t)e^{-st}\, dt = \lim_{b \to \infty} \int_0^b f'(t)e^{-st}\, dt \\ &= \lim_{b \to \infty} \left[f(t)e^{-st}\big|_0^b - \int_0^b f(t)(-se^{-st})\, dt \right] \\ &= -f(0) + s \lim_{b \to \infty} \int_0^b f(t)e^{-st}\, dt \\ &= s\mathcal{L}(f)(s) - f(0). \end{aligned} \tag{11.1}$$

Here we have assumed that f is differentiable and f' is piecewise continuous and of exponential order. We can summarize (11.1) as follows: If the Laplace transform of f is $F(s)$, then the Laplace transform of f' is $sF(s) - f(0)$. In other words, the Laplace transform changes the operation of differentiation into the operation of multiplication (by the independent variable) plus a translation (by $-f(0)$).

Remark 11.1 If you know $\mathcal{L}(f')(s)$ and $f(0)$, you can add them and divide by s to obtain $\mathcal{L}(f)(s)$. This procedure is analogous to integration.

Applying the formula in (11.1) repeatedly yields the following generalization to higher derivatives. If the Laplace transform of f is F, $f^{(k)}$ is continuous for $k = 0, \ldots, n-1$, and $f^{(n)}$ is piecewise continuous, then

$$\mathcal{L}(f^{(n)})(s) = s^n F(s) - s^{n-1}f(0) - s^{n-2}f'(0) - \cdots - f^{(n-1)}(0).$$

The Laplace transform has many other important properties, of which we mention three here. First, the Laplace transform is invertible, in the sense that knowing the function $\mathcal{L}(f)(s)$ determines $f(t)$ for $t \geq 0$, except at points where it is discontinuous. (This is as good as we can expect, since changing f for $t < 0$ does not affect $\mathcal{L}(f)$, nor does changing its value at a point of discontinuity.) The inverse Laplace transform is denoted $\mathcal{L}^{-1}$. Like the Laplace transform, the inverse Laplace transform can be written as an integral transform, but it involves contour integrals in the complex plane, so we do not give the definition here. Second, the Laplace transform is linear; *i.e.*, $\mathcal{L}(af + bg) = a\mathcal{L}(f) + b\mathcal{L}(g)$, where a and b are constants. Linearity of the Laplace transform follows easily from linearity of integration. Third, the inverse Laplace transform is linear. Linearity of the inverse Laplace transform follows from the linearity of the Laplace transform.

11.1 Differential Equations and Laplace Transforms

Let's see what happens when we apply the Laplace transform to a second order linear differential equation with constant coefficients. Consider the initial value problem

$$ay''(t) + by'(t) + cy(t) = f(t), \qquad y(0) = y_0, \qquad y'(0) = y_0'.$$

The function $f(t)$ is called the forcing function of the differential equation, because in many physical models $f(t)$ corresponds to the influence of an external force. If we apply the Laplace transform to this equation and use the initial conditions, we get the algebraic equation

$$a(s^2 Y(s) - sy_0 - y_0') + b(sY(s) - y_0) + cY(s) = F(s),$$

where $Y(s)$ is the Laplace transform of $y(t)$ and $F(s)$ is the Laplace transform of $f(t)$. We solve this algebraic equation for $Y(s)$ to get

$$Y(s) = \frac{F(s) + asy_0 + ay_0' + by_0}{as^2 + bs + c},$$

and then compute the inverse Laplace transform of the right-hand side to get an expression for $y(t)$. The resulting solution is only guaranteed to be valid for $t \geq 0$, since the Laplace transform method only takes into account how $f(t)$ behaves for $t \geq 0$.

Remark 11.2 In control theory, a common problem is to find a forcing function $f(t)$ that will make the solution $y(t)$ behave in a desired fashion. Having taken the Laplace transform of the differential equation, this problem becomes algebraic: find $F(s)$ that makes $Y(s)$ have a desired form. One must work with Laplace transforms for a while to develop an understanding of how the form of $Y(s)$ determines the behavior of the solution $y(t)$. But having developed this sense, many engineers and scientists find it helpful to think in terms of the Laplace transform of a differential equation that models a system rather than think directly about the differential equation. While the equations we consider in this chapter can be solved directly using **DSolve**, one purpose of the examples in this section and the corresponding problems in Problem Set E is for you to see how the solutions and their Laplace transforms correspond.

Traditionally, one used tables to look up Laplace transforms and inverse Laplace transforms. But you can use the *Mathematica* commands **LaplaceTransform** and **InverseLaplaceTransform** to compute Laplace transforms and inverse Laplace transforms. For example, to compute the Laplace transform of $\cos t$, type:

```
LaplaceTransform[Cos[t], t, s]
```

$$\frac{s}{1+s^2}$$

The second and third input arguments to **LaplaceTransform** specify the independent variables of the original function and the transformed function. To compute the inverse Laplace transform of the output above, type:

```
InverseLaplaceTransform[s/(1+s^2), s, t]
```

```
Cos[t]
```

Exercise 11.2 Compute the Laplace transform of t^2 and e^{at}. Compute the inverse Laplace transform of $1/(s^2 - 1)$.

We can use these commands to implement the Laplace transform method for solving differential equations.

Example 11.1 Consider the initial value problem $y'' + y = \sin 2t$, $y(0) = 1$, $y'(0) = 0$. First, we define the equation that we want to solve and compute its Laplace transform. It is convenient to replace the lengthy expression

```
LaplaceTransform[y[t], t, s]
```

by **Y[s]**, since it occurs several times in the output.

```
eqn1 = y''[t] + y[t] == Sin[2 t];
lteqn = LaplaceTransform[eqn1, t, s]
        /. LaplaceTransform[y[t], t, s] → Y[s]
```

$$-s \; y[0] + Y[s] + s^2 \; Y[s] - y'[0] == \frac{2}{4+s^2}$$

Now we solve this equation for the Laplace transform of $y(t)$, assuming the initial conditions $y(0) = 1$, $y'(0) = 0$.

```
ltsol = Solve[lteqn, y[0] == 1, y'[0] == 0, Y[s]]
```

$$\{\{ \; Y[s] \; \rightarrow \; \frac{s + \dfrac{2}{4+s^2}}{1+s^2} \}\}$$

Finally, take the inverse Laplace transform:

```
ysol = InverseLaplaceTransform[Y[s] /. First[ltsol], s, t]
```

$$\frac{1}{3} \; (3 \; \text{Cos}[t] + 2 \; \text{Sin}[t] - \text{Sin}[2 \; t])$$

You can now evaluate or plot this solution. For example, to plot the solution over the interval $[0, 5]$, type:

```
Plot[ysol, {t, 0, 5}, PlotRange → All]
```

The resulting graph appears in Figure 11.1.

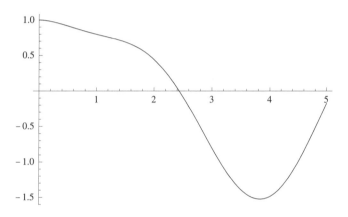

Figure 11.1: Laplace Transform Solution of $y'' + y = \sin 2t$, $y(0) = 1$, $y'(0) = 0$

By following the steps above you can solve other linear equations with constant co-efficients. The only things you need to change are the definition of the equation and the values you substitute for the initial conditions. We can automate this process by defining a procedure **LTSolve** as follows:

```
LTSolve[eqn_, y0_, yp0_] := Module[{lteqn, ltsol, s, Y},
        lteqn = LaplaceTransform[eqn, t, s]
                /. LaplaceTransform[y[t], t, s] → Y[s];
        ltsol = Solve[{lteqn, y0, yp0}, Y[s]];
        InverseLaplaceTransform[Y[s] /. First[ltsol], s, t]]
```

Sure enough, it works:

```
LTSolve[eqn1, y[0] == 1, y'[0] == 0]
```

$$\frac{1}{3} \; (3 \; \text{Cos[t]} \; + \; 2 \; \text{Sin[t]} \; - \; \text{Sin[2 t]})$$

Example 11.2 Consider the initial value problem:

$$y''(t) + 2y'(t) = e^{-t}, \quad y(0) = 1, \quad y'(0) = 2.$$

Here are the simple *Mathematica* commands for producing a solution.

```
eqn2 = y''[t] + 2 y'[t] == Exp[-t];
LTSolve[eqn2, y[0] == 1, y'[0] == 2]
```

$$\frac{5}{2} - \frac{e^{-2\,t}}{2} - e^{-t}$$

11.2   Discontinuous Functions

The Laplace transform is especially useful for solving differential equations that involve piecewise continuous functions. The basic building block for piecewise continuous functions is the *unit step function* $u(t)$, defined by

$$u(t) = \begin{cases} 0, & \text{if } t < 0, \\ 1, & \text{if } t \geq 0. \end{cases}$$

In honor of Oliver Heaviside (1850–1925), who developed the Laplace transform method to solve problems in electrical engineering, this function is sometimes called the Heaviside function. In *Mathematica* it is called **UnitStep** or **HeavisideTheta**. Strictly speaking, these two *Mathematica* functions are not the same. **UnitStep** is defined as above to have the value 1 at 0; whereas **HeavisideTheta** is also defined as above, *except* that at 0, it is left undefined. But both of these functions have the same Laplace transform, since they differ only at a single point.

The unit step function is best thought of as a switch, which is off until time 0 and then is on starting at time 0. To make a switch that comes on at time c, we simply translate by c; thus $u(t - c)$ is a switch that comes on at time c. Similarly, $1 - u(t - c)$ is a switch that goes off at time c. The function $u(t - c)$ is sometimes written $u_c(t)$.

The unit step function can be used to build piecewise continuous functions by switching pieces of the function on and off at appropriate times. Consider, for example, the function

$$f(t) = \begin{cases} 0, & \text{if } t < 0, \\ 1, & \text{if } 0 \leq t < 1, \\ t^2, & \text{if } 1 \leq t < 3, \\ \sin 2t, & \text{if } t \geq 3. \end{cases}$$

We can write this as

$$f(t) = 0 + u(t)(1 - 0) + u(t - 1)(t^2 - 1) + u(t - 3)(\sin 2t - t^2).$$

We started with the formula for $f(t)$, in this case 0, that is valid for negative values of t. Then for each point c where the formula for $f(t)$ changes, in this case $c = 0, 1, 3$, we added the term

$$u(t - c)(\text{formula to the right of } c - \text{formula to the left of } c).$$

In other words, at each value of c we switched on the difference between the formula for $f(t)$ that is valid just to the right of c and the formula that is valid just to the left of c. In *Mathematica*, you can enter $f(t)$ as

```
f[t_] = UnitStep[t] + UnitStep[t-1]*(t^2-1)
        + UnitStep[t-3]*(Sin[2t]-t^2)
```

$\left(-t^2+\text{Sin}[2\ t]\right)\ \text{UnitStep}[-3+t]+\left(-1+t^2\right)\ \text{UnitStep}[-1+t]+\text{UnitStep}[t]$

To plot $f(t)$, type:

```
Plot[f[t], {t, 0, 10}, PlotRange → All]
```

The result appears in Figure 11.2. In this graph, you can clearly see the three different "pieces" of the function.

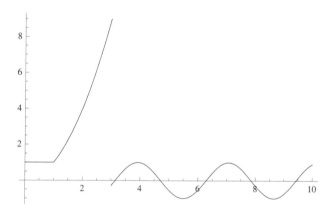

Figure 11.2: A Piecewise Continuous Function

You can also evaluate $f(t)$ at any point:

```
f[2]
```

4

On the other hand, **UnitStep** and **HeavisideTheta** differ at 0, as we can see from:

```
{UnitStep[0], HeavisideTheta[0]}
```

$\{1, \text{HeavisideTheta}[0]\}$

Another important discontinuous function is the *Dirac delta function*, usually denoted $\delta(t)$, and sometimes called the *unit impulse function*. The delta function belongs to a class of mathematical objects called *generalized functions* (it is not a true function). By definition, $\delta(t) = 0$ for $t \neq 0$, but $\int_{-\infty}^{\infty} \delta(t)dt = 1$. There is no finite value of $\delta(0)$ that can make its integral nonzero, so we must regard $\delta(0)$ as infinite.

The motivation for the Dirac delta function is as follows. In physics, the total impulse imparted by a varying force $F(t)$ is the integral of $F(t)$. If we consider a force of total impulse 1 and make this force act over a smaller and smaller time interval around 0, then in the limit we obtain $\delta(t)$. Thus the delta function represents an idealized force of total impulse 1 concentrated at the instant $t = 0$. The magnitude and timing of the impulse can be changed by multiplying the delta function by a constant and translating it. Thus the function $10\delta(t - 8)$ represents a force of total impulse 10 concentrated at the instant $t = 8$. The function $\delta(t - c)$ is sometimes denoted $\delta_c(t)$. In *Mathematica*, the Dirac delta function is called `DiracDelta`.

Exercise 11.3 Plot the Dirac delta function on the interval $[-1, 1]$ (you won't see the impulse at 0). Evaluate the function at 0 and 1. Integrate the function over the interval $[-1, 1]$.

We consider now the Laplace transforms of $u_c(t)$ and $\delta_c(t)$. Since the Laplace transform of a function depends only on its values for $t \geq 0$, we consider only the case when $c \geq 0$ (otherwise these functions are just 1 and 0, respectively, for $t \geq 0$). Then

$$\mathcal{L}(u_c)(s) = \int_0^{\infty} u_c(t)e^{-st}\, dt = \int_c^{\infty} e^{-st}\, dt = \frac{e^{-cs}}{s}, \quad \text{for } s > 0.$$

Strictly speaking, taking the Laplace transform of $\delta_c(t)$ requires the theory of generalized functions, which we alluded to above. Here is a rough explanation. First, we think of $\delta_c(t)$ as a limit of *bona fide* functions. The easiest way to do this is to use step functions. We can write a function of total integral 1, concentrated on the interval $[c - \varepsilon, c + \varepsilon]$, as

$$\frac{1}{2\varepsilon}(u_{c-\varepsilon}(t) - u_{c+\varepsilon}(t)) = \frac{1}{2\varepsilon}(u(t - c + \varepsilon) - u(t - c - \varepsilon)).$$

Thus in a certain sense,

$$\delta_c(t) = \lim_{\varepsilon \to 0^+} \frac{1}{2\varepsilon}(u(t - c + \varepsilon) - u(t - c - \varepsilon)) = u'(t - c) = u'_c(t).$$

The rule (11.1) for the Laplace transform of a derivative can be justified in the case that the derivative is a generalized function, so for $c > 0$,

$$\mathcal{L}(\delta_c)(s) = s\mathcal{L}(u_c)(s) - u_c(0) = e^{-cs}.$$

Exercise 11.4 Verify that `LaplaceTransform` correctly computes the Laplace transforms of the functions `DiracDelta[t-3]`, `HeavisideTheta[t-2]`, and `UnitStep[t-2]`.

11.3 Differential Equations with Discontinuous Forcing

Consider an inhomogeneous, second order linear equation with constant coefficients:

$$ay''(t) + by'(t) + cy(t) = g(t).$$

If the forcing function $g(t)$ is piecewise continuous or involves the delta function, then we can solve the equation by the method of Laplace transforms.

Example 11.3 Consider the initial value problem:

$$y''(t) + 3y'(t) + y(t) = g(t), \quad y(0) = 1, \quad y'(0) = 1,$$

where

$$g(t) = \begin{cases} 0, & \text{if } t < 0, \\ 1, & \text{if } 0 \leq t < 1, \\ -1, & \text{if } 1 \leq t < 2, \\ 0, & \text{if } t \geq 2. \end{cases}$$

We can solve this equation in *Mathematica* with the following sequence of commands:

```
g[t_] = UnitStep[t] - 2 UnitStep[t-1] + UnitStep[t-2];
eqn3 = y''[t] + 3y'[t] + y[t] == g[t];
sol3 = LTSolve[eqn3, y[0] == 1, y'[0] == 1]
```

The formula for the solution is very long, so we omit it here. To graph the solution, type:

```
Plot[sol3, {t, 0, 5}, PlotRange → All]
```

Figure 11.3 shows the graph.

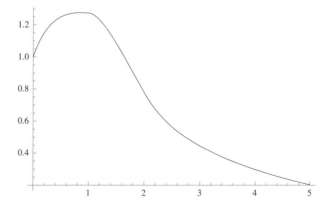

Figure 11.3: Solution of an Equation with Piecewise Continuous Data

Even though the forcing function $g(t)$ is discontinuous at $t = 2$, the solution $y(t)$ is a continuous function. The same is true of solutions of any second order linear differential equation with a discontinuous forcing function. In fact, as long as the forcing function is bounded, the solutions are differentiable too. If the forcing function involves the Dirac delta function, the solution will be continuous but not differentiable at the times of the impulses, as we will see in the next example.

Example 11.4 We can also use *Mathematica* to solve inhomogeneous equations involving the delta function. Consider the initial value problem

$$y''(t) + y'(t) + y(t) = \delta_1(t), \quad y(0) = 0, \quad y'(0) = 0.$$

We can solve this equation with the following sequence of commands:

```
eqn4 = y''[t] + y'[t] + y[t] == DiracDelta[t-1];
sol4 = LTSolve[eqn4, y[0] == 0, y'[0] == 0]
```

$$\frac{2\, e^{\frac{1-t}{2}}\, \text{HeavisideTheta}[-1+t]\, \text{Sin}\!\left[\frac{1}{2}\sqrt{3}\,(-1+t)\right]}{\sqrt{3}}$$

Figure 11.4 contains a graph of the solution. In this graph, you can clearly see the effect of the unit impulse at time $t = 1$. Until $t = 1$, the solution is 0. At time $t = 1$, the unit impulse instantaneously changes the velocity y' to 1, though the solution remains continuous. After $t = 1$, the impulse function has no further effect; it is as if we solved the homogeneous equation $y'' + y' + y = 0$ with initial conditions $y(1) = 0$ and $y'(1) = 1$ for $t \geq 1$. Since the roots of the characteristic polynomial are complex with negative real parts, the solution decays to 0 in an oscillatory manner.

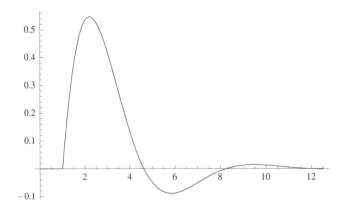

Figure 11.4: Solution of an Equation with an Impulse

Problem Set E

Series Solutions and Laplace Transforms

Problems 1–8 concern series solutions, whereas Problems 9–20 concern Laplace transforms and/or equations with discontinuous forcing functions. The solution to Problem 1 appears in the *Sample Solutions*.

1. Consider Airy's equation

$$y'' - xy = 0.$$

(a) Compute the terms of degree 10 or less in the Taylor series expansion of the solution $y(x)$ to Airy's equation $y'' - xy = 0$ that satisfies $y(0) = 1$, $y'(0) = 0$.

(b) On the intervals $(0, 5)$ and $(-10, 0)$, graph the Taylor polynomial you obtained, together with the exact solution of the initial value problem. (You can compute the exact solution with **DSolve**. The solution will be given in terms of the special functions **AiryAi**, **AiryBi**, and **Gamma**.) Where is the approximation accurate? (*Hint*: You may have to restrict the range on the plot over negative values.)

(c) Airy's equation can be written as $y'' = xy$. If $x > 1$, then $y'' = xy > y$, so the solutions of Airy's equation grow at least as quickly as the solutions of $y'' = y$. Analyze the latter equation, and predict how well Taylor polynomials will approximate exact solutions of Airy's equation. Relate your analysis to the graph in part (b).

(d) Recall from Problem 1 in Problem Set D that the behavior of the solutions to Airy's equation for negative values is oscillatory. Relate that to your graph in part (b) and explain why it fails to be accurate beyond a certain point.

2. Compute the terms of degree 10 or less in the Taylor series expansion of the solution to the following differential equation (*cf*. Problem 23, Section 5.2 of Boyce and DiPrima):

$$y'' - xy' - y = 0, \quad y(0) = 1, \quad y'(0) = 0. \tag{E.1}$$

(a) Calculate and graph the series solution—that is, the Taylor polynomial—of degree 10.

(b) What is the long-term behavior of the Taylor polynomial (as $x \to \pm\infty$)? Is the series solution even, odd, or neither?

(c) Find the general solution of the differential equation in (E.1) using **DSolve**, and then find values of the constants so that the initial conditions in (E.1) are satisfied. What is the formula for the solution of (E.1)? Are your conclusions in part (b) confirmed by this formula?

(d) Use the command **Series** to calculate the Taylor expansion of the formula obtained in part (c). Does this Taylor expansion agree with the series solution in part (a)?

3. Consider the initial value problem (*cf.* Problem 18, Section 5.2 of Boyce and DiPrima):

$$(1 - x)y'' + xy' - y = 0, \quad y(0) = -3, \quad y'(0) = 2.$$

(a) Compute the terms of degree 10 or less in the Taylor series expansion of the solution.

(b) The theory suggests that the true solution may have a singularity at $x = 1$. Graph the Taylor polynomial you obtained. Do you see any signs of the singularity?

(c) Compute the exact solution using **DSolve**. Does the solution have a singularity?

(d) Compute the general solution of the differential equation and discuss its behavior at $x = 1$.

(e) Find the first few terms in the Taylor series expansion at $x = 1$ for two linearly independent solutions of the equation.

4. Use **DSolve** to solve the following Euler equations (*cf.* Section 5.4 of Boyce and DiPrima, Problems $3, 6, 8, 14$):

(a) $x^2 y'' - 3xy' + 4y = 0$,

(b) $(x - 1)^2 y'' + 8(x - 1)y' + 12y = 0$,

(c) $2x^2 y'' - 4xy' + 6y = 0$,

(d) $4x^2 y'' + 8xy' + 17y = 0$, $y(1) = 2$, $y'(1) = -3$. In this part, graph the solution and describe its behavior as $x \to 0$.

5. For the following differential equations, compute the indicial equation and find its roots r_1 and r_2. Then compute the terms of degree 10 or less in the Frobenius series solution corresponding to the larger root. If $r_1 - r_2$ is not an integer, do the same for the other root (*cf.* Problems $1, 5, 8$ of Section 5.5 of Boyce and DiPrima):

(a) $2xy'' + y' + xy = 0$,

(b) $3x^2y'' + 2xy' + x^2y = 0$,

(c) $2x^2y'' + 3xy' + (2x^2 - 1)y = 0$.

6. This problem concerns Bessel's equation, which arises in many physical problems with circular symmetry, such as the study of water waves in a circular pond, vibrations of a circular drum, or the diffraction of light by the circular aperture of a telescope. The Bessel function $J_n(x)$ (where the n is an arbitrary integer) is defined to be the coefficient of t^n in the power series expansion of the function

$$\exp\left(\frac{1}{2}x\left(t - \frac{1}{t}\right)\right).$$

(By thinking about what happens to the power series when t and $1/t$ are switched, you can see that $J_{-n}(x) = J_n(x)$ for n even and $J_{-n}(x) = -J_n(x)$ for n odd.) $J_n(x)$ is also a solution to *Bessel's equation of order* n,

$$x^2y'' + xy' + (x^2 - n^2)y = 0.$$

The Taylor series expansion for J_0 is (*cf.* Section 5.7 of Boyce and DiPrima):

$$J_0(x) = \sum_{j=0}^{\infty}(-1)^j \frac{1}{2^{2j}} \frac{x^{2j}}{(j!)^2}.$$

(a) *Mathematica* has a built-in function for $J_n(x)$, denoted by **BesselJ[n,x]**. Use it to plot $J_0(x)$ for $0 \le x \le 10$. (You don't need to graph the function for negative x since J_0 is an even function.) On the same axes, draw the graphs of the 4th order, 10th order, and 20th order Taylor polynomials of J_0. How well do the Taylor polynomials approximate the function?

(b) An interesting identity satisfied by the Bessel functions is

$$J_n'(x) = \frac{1}{2}\left(J_{n-1}(x) - J_{n+1}(x)\right).$$

In particular,

$$J_0'(x) = \frac{1}{2}\left(J_{-1}(x) - J_1(x)\right) = \frac{1}{2}\left(-J_1(x) - J_1(x)\right) = -J_1(x).$$

Thus the maxima and minima of J_0 occur at the zeros of J_1. Using **FindRoot**, solve the equations $J_0(x) = 0$ and $J_1(x) = 0$ numerically to find the first five zeros and the first five relative extreme points of J_0, starting at $x = 0$. (You will need 10 different starting intervals—use your graph for guidance.) Compute the differences between successive zeros. Can you make a guess about the periods of the oscillations as $x \to \infty$? (If you did Problem 15 in Problem Set D, you know exactly how the period behaves as $x \to \infty$.)

(c) From the general theory, one knows that Bessel's equation of order n has another linearly independent solution $Y_n(x)$, with a logarithmic singularity as $x \to 0^+$. *Mathematica* also has a built-in function, `BesselY[n,x]`, for computing $Y_n(x)$. Graph the function $Y_0(x)$ on the interval $0 < x \le 10$. Then compute $c = \lim_{x \to 0+} Y_0(x) / \ln x$. Graph the function

$$Y_0(x) - c \ln x$$

on the same interval, and observe that the function has a limit as x approaches 0 from the right. Thus the "singular part" of $Y_0(x)$ behaves like $c \ln x$.

7. The first order homogeneous linear differential equation

$$x^2 y' + y = 0$$

has an irregular singular point at $x = 0$ (why?); however, as a first order linear equation it can be solved exactly.

(a) Find the general solution of the equation for $x > 0$ and for $x < 0$. How do the solutions behave as $x \to 0^+$? as $x \to 0^-$? Use *Mathematica* to graph a solution for $0 < x < 2$ and a solution for $-2 < x < 0$.

(b) Now find (for $x > 0$ and for $x < 0$) the general solution of the Euler equation

$$xy' + ry = 0,$$

which has a regular singular point at $x = 0$ if $r \ne 0$. How do the solutions behave as $x \to 0^+$? as $x \to 0^-$? Use *Mathematica* to graph solutions to the right and left of 0 for each of the cases $r = 0.5, r = 1, r = -0.5, r = -1$. (Be careful! If you compute a general solution, then you will have to allow complex values of the undetermined constant to get all the real solutions.)

(c) What differences do you notice between the behavior of solutions near a regular singular point and the behavior near an irregular singular point?

8. Nonlinear differential equations can sometimes be solved by series methods. Consider, for example,

$$y' = y^2 + x^2. \tag{E.2}$$

Let's use the initial condition $y(0) = 0$, and look for a solution of the form

$$y(x) = \sum_{i=1}^{\infty} a_i x^i.$$

(We know there is no constant term in the Taylor series expansion of y at $x = 0$ because of the initial condition.)

(a) Explain why a_1 must equal 0.

(b) Now use the scheme from Chapter 10 to compute a Taylor series expansion of the solution to (E.2). Make sure you go far enough to get the 11th degree monomial. What pattern do you see?

(c) In particular, only odd powers of x appear. Show that the solution to the IVP $y' = y^2 + x^2$, $y'(0) = 0$ must be an odd function, that is, $y(-x) = -y(x)$. (*Hint:* Set $w(x) = -y(-x)$ and show that w solves the IVP.)

(d) Plot the 11th degree Taylor polynomial for y, together with a numerical solution of (E.2) obtained using **NDSolve**, on the interval $-1 < x < 1$. How close are the graphs? Now try plotting on the interval $-1.95 < x < 1.95$. What is happening? Next note that for $x > 0$, we have $y'(x) > x^2$, so $y(x) > x^3/3$. In particular, $y(1) > 1/3$. But for $x > 1$, the right-hand side of (E.2) is greater than $y(x)^2 + 1$; so the solution grows faster than the solution to the IVP $y' = y^2 + 1$, $y(1) = 1/3$. Use **DSolve** to solve that IVP. You get a solution of the form $y_1(x) = \tan(x - C)$. Using the fact that the tangent function blows up when its argument reaches $\pi/2$, find the value v_1 at which y_1 blows up. (It will be approximately 2.2.) Hence the solution of (E.2) must become infinite somewhere between $x = 1$ and $x = v_1$. Is there any way of seeing this from the series solution? Use **NDSolve** to locate the point P where the solution blows up. Graph the two approximate solutions (that is, the series solution out to degree 11 and the numerical solution) on the interval $(0, P - \epsilon)$, where you will have to pick the small positive number ϵ carefully to get a useful plot. What do you observe?

9. For each of the following functions, compute the Laplace transform. Then plot the Laplace transform over the interval $[0, 10]$.

(a) $\sin t$.

(b) e^t.

(c) $\cos t$.

(d) $t \cos t$.

(e) $u_1(t) \sin t$.

On the basis of these graphs, what general conclusions can you draw about the Laplace transform of a function? In particular, what can you say about the growth or decay of the Laplace transform of a function? Can you justify your conclusion by looking at the integral formula for the Laplace transform? (*Hint:* Look at the derivation of $\mathcal{L}(e^{at})$ at the beginning of Chapter 11.)

10. Suppose $f(t)$ and $g(t)$ are functions with Laplace transforms $\mathcal{L}(f) = F$ and $\mathcal{L}(g) = G$. Verify the following identities with *Mathematica*. You might find it useful to replace the **LaplaceTransform** command with something like:

```
LTSimplified[fcn_,t_,s_] = LaplaceTransform[fcn,t,s]
   /. {LaplaceTransform[f[t], t, s]  →  F[s],
        LaplaceTransform[g[t], t, s]  →  G[s]}
```

(a) $\mathcal{L}(af + bg) = a\mathcal{L}(f) + b\mathcal{L}(g)$, where a and b are real numbers.

(b) $\mathcal{L}(e^{ct}f(t)) = F(s - c)$.

(c) $\mathcal{L}(f') = sF(s) - f(0)$.

(d) $\mathcal{L}(\int_0^t f(t - u)g(u)\, du) = F(s)G(s)$.

(e) $\mathcal{L}(u_1(t)f(t - 1)) = e^{-s}F(s)$.

The function $t \to \int_0^t f(t - u)g(u)\, du$ is called the *convolution* of f and g, and is written $f * g$. Thus part (d) says that the Laplace transform changes convolution into ordinary multiplication. As a hint for part (e), specialize (d) to the case where $g(t)$ is $\delta(t - 1)$.

11. Consider the following IVP:

$$y'' + 2y' + 2y = g(t),\ \ y(0) = 0,\ \ y'(0) = 0,\ \ g(t) = \begin{cases} \sin t, & \text{if } 0 \le t < \pi \\ 0, & \text{if } t \ge \pi. \end{cases} \quad \text{(E.3)}$$

(a) Solve the equation for $0 \le t \le \pi$ by using **DSolve** on the IVP

$$y'' + 2y' + 2y = \sin t, \quad y(0) = 0, \quad y'(0) = 0.$$

Plot the solution on the interval $0 \le t \le \pi$.

(b) Find numerical values for $y_\pi = y(\pi)$ and $y'_\pi = y'(\pi)$ using the solution from part (a). Then use **DSolve** to solve the IVP

$$y'' + 2y' + 2y = 0, \quad y(\pi) = y_\pi, \quad y'(\pi) = y'_\pi.$$

Plot the solution on the interval $\pi \le t \le 15$. Combine this graph with your graph from part (a) to show the solution to (E.3) on the entire interval $0 \le t \le 15$.

(c) Now solve the equation using the Laplace transform method. Plot the solution on $[0, 15]$, and compare it with the solution from part (b).

(d) Find the general solution of the associated homogeneous equation. What is the asymptotic behavior of those solutions as $t \to \infty$? Knowing that the forcing term of the inhomogeneous equation is zero after $t = \pi$, what can you say about the long-term behavior of the solution of the inhomogeneous equation?

12. Use the Laplace transform method to solve the following initial value problems. Then graph the solutions on the interval $[0, 15]$. In parts (a) and (b), also graph the function on the right-hand side of the differential equation (the forcing function). In each part, discuss how the solution behaves in response to the forcing function.

(a) $y'' + 3y' + 2y = h(t), \quad y(0) = 1, \quad y'(0) = 0,$

$$h(t) = \begin{cases} 1, & \text{if } 0 \le t \le 10 \\ 0, & \text{if } t > 10. \end{cases}$$

(b) $y'' + 2y' + 3y = u_3(t), \quad y(0) = 0, \quad y'(0) = 1.$

(c) $y'' + 2y' + \dfrac{4}{5}y = g(t), \quad y(0) = 0, \quad y'(0) = 0,$

$$g(t) = \begin{cases} \cos(t), & \text{if } 0 \le t \le \pi \\ 0, & \text{if } t > \pi. \end{cases}$$

(d) $y'' + 4y = \delta(t - \pi) - \delta(t - 2\pi), \quad y(0) = 0, \quad y'(0) = 0.$

(e) $y'' + 2y' + 3y = \sin t + \delta(t - 3\pi), \quad y(0) = 0, \quad y'(0) = 0.$

13. Use the Laplace transform method to solve the following initial value problems. See Problem 12 for additional instructions.

(a) $y'' + 4y = \sin t - u_{2\pi}(t)\sin(t - 2\pi), \quad y(0) = 0, \quad y'(0) = 0.$

(b) $y'' + 6y' + 8y = h(t), \quad y(0) = 0, \quad y'(0) = 2, \quad h(t) = \begin{cases} 0, & 0 \le t < 5 \\ 1, & 5 \le t < 10 \\ -1, & t \ge 10. \end{cases}$

(c) $y'' + 4y = \delta(t - 3\pi), \quad y(0) = 1, \quad y'(0) = 0.$

(d) $y'' + y = \delta(t - 2) - \delta(t - 8), \quad y(0) = 0, \quad y'(0) = 0.$

14. Use the Laplace transform method to solve the following initial value problems of higher order. Once you have solved the equation, plot the solution on an appropriate interval.

(a) $y'''(t) - y''(t) - y'(t) - 2y(t) = \delta(t - 1), \quad y(0) = y'(0) = y''(0) = 0.$

(b) $y^{(4)}(t) + 2y''(t) + y(t) = \cos t, \quad y(0) = 1, \quad y'(0) = y''(0) = y'''(0) = 0.$

(c) $y^{(4)}(t) + 3y''(t) - 4y(t) = u_1(t)\sin(2t), \quad y(0) = y'(0) = y''(0) = y'''(0) = 0.$

(d) $y'''(t) + 4y'(t) = \delta(t - \pi), \quad y(0) = y''(0) = 0, \quad y'(0) = 1.$

Which of these equations has resonance-type behavior?

15. This problem is based on Problem 35 in Section 6.2 of Boyce and DiPrima. Consider Bessel's equation

$$ty'' + y' + ty = 0.$$

We will use the Laplace transform to find some leading terms of the power series expansion of the Bessel function of order zero that is continuous at the origin. Let $y(t)$ be such a solution of the equation, and suppose $Y(s)$ is its transform.

(a) Assume that $y(0) = 1$, $y'(0) = 0$. Show that $Y(s)$ satisfies the equation

$$(1 + s^2)Y'(s) + sY(s) = 0.$$

You will have to do this in steps. First compute

```
LaplaceTransform[y[t]*t,t,s] +
    D[LaplaceTransform[y[t],t,s],s]
```

to show that $\mathcal{L}(t\,y(t))$ is $-Y'(s)$. Then apply this with y'' in place of y and put the results together.

(b) Solve the equation in part (a) using **DSolve**. (Use $Y(0) = 1$.)

(c) Use **Series** to expand the solution in part (b) in powers of $1/s$ out to terms of degree 11. (*Hint*: The expansion should be valid as $s \to \infty$; if you ask **Series** to expand about $s = \infty$, you will get a series in powers of $1/s$.)

(d) Apply the inverse Laplace transform to part (c) to obtain the power series expansion of y in powers of t. To what degree is this power series valid?

16. This problem illustrates how the choice of method can dramatically affect the time it takes the computer to solve a differential equation. It can also affect the form of the solution. Consider the initial value problem

$$y'' + y' + y = (t+1)^3 e^{-t} \cos t, \qquad y(0) = 1, \qquad y'(0) = 0. \qquad \text{(E.4)}$$

This problem could be solved by the methods of undetermined coefficients, variation of parameters, or Laplace transforms.

(a) Use **Timing[DSolve[...]]** to solve the problem. This will give not only the result, but also the time it takes *Mathematica* to do the calculation. By default, *Mathematica* selects a solution method that turns out to be quite inefficient; hence the answer comes out in a complicated form.

(b) Now use **Timing** and the Laplace transform method to solve the problem. Compare the time used with this method with the time from part (a).

(c) Verify that the expression produced in part (b) really is a solution to the differential equation. Plot the solution over the range $0 \le t \le 15$.

17. In this problem we investigate the effect of a periodic discontinuous forcing function (a square wave) on a second order linear equation with constant coefficients. Consider the initial value problem

$$y'' + y = h(t), \quad y(0) = 0, \ y'(0) = 1. \qquad \text{(E.5)}$$

The associated homogeneous equation has natural period 2π; the general solution of the homogeneous equation is

$$y(t) = A\cos t + B\sin t.$$

Recall that the phenomenon of resonance occurs when the forcing function $h(t)$ is a linear combination of $\sin t$ and $\cos t$. Does resonance occur when the forcing function is periodic of period 2π but discontinuous?

(a) Using step functions, define a *Mathematica* function $h(t)$ on the interval $[0, 10\pi]$ whose value is $+1$ on $[0, \pi)$, -1 on $[\pi, 2\pi)$, $+1$ on $[2\pi, 3\pi)$, and so on. Plot the function on the interval $[0, 30]$. It should have the appearance of a square wave.

(b) Use the Laplace transform method from Chapter 11 to solve equation (E.5) with the function $h(t)$ defined in part (a). Plot the solution together with $h(t)$ on the interval $[0, 30]$. Do you see resonance?

(c) In part (a), we constructed a forcing function $h(t)$ with period 2π. The function $h(t/2)$ has period 4π. Repeat part (b) using the forcing function $h(t/2)$. Do you see resonance?

(d) Repeat part (b) using the forcing function $h(2t)$; this time just plot from 0 to 15. Do you see resonance? What is the period of $h(2t)$?

(e) What can you conclude about the resonance effect for discontinuous forcing functions? Would you expect resonance to occur in equation (E.5) for *any* forcing function of period 2π? (*Hint:* The function $h(2t)$ has period π as well as period 2π.) Can you venture a guess about when resonance occurs for discontinuous periodic forcing functions? You might try some other periodic forcing functions to check your guess.

18. One additional use of the Laplace transform is to transform a *partial differential equation*, involving both time and space derivatives, into an equation involving only space derivatives. We will illustrate this process with the *heat equation* in one space variable, which models the temperature $u(x, t)$ in a rod as a function of position x and time t. This equation takes the form

$$\frac{\partial u}{\partial t} = k\frac{\partial^2 u}{\partial x^2}, \tag{E.6}$$

where k is a constant depending on the material of the rod. For simplicity, we will assume units have been chosen so that $k = 1$. Let's assume our rod has length 1 (again, in suitable units) and that the temperature at the two ends of the rod is kept fixed, say at $u = 0$. Then we have a *boundary value problem* of the form

$$u(0, t) = u(1, t) = 0, \qquad u(x, 0) = u_0(x), \tag{E.7}$$

where u_0 is the initial temperature distribution in the rod at $t = 0$. Note that u_0 should satisfy $u_0(0) = u_0(1) = 0$, for compatibility with (E.7).

We can solve this problem as follows. Take the Laplace transform $U(x, s)$ of $u(x, t)$ in t, thinking of x as an extra parameter. Then (E.6) and (E.7) (with $k = 1$) become a boundary value problem for an ordinary differential equation for U, with x as the independent variable and s as an extra parameter:

$$sU(x, s) - u_0(x) = \frac{d^2U(x, s)}{dx^2}, \qquad U(0, s) = U(1, s) = 0. \qquad \text{(E.8)}$$

We can solve (E.8), say with **DSolve**, and then take the inverse Laplace transform to recover the function u.

(a) Carry out this process to solve (E.6) and (E.7) (with $k = 1$) when the initial temperature is $u_0(x) = \sin(\pi x)$. You should get a nice simple formula for the answer; check that it indeed satisfies (E.6).

(b) Repeat part (a) with $u_0(x) = \sin(n\pi x)$, $n = 2, 3, 4$. (You can do this with a loop.) You should see a general pattern! What is the solution for general n? Check it! (Note: It doesn't work simply to try to do the general case all at once, because *Mathematica* doesn't know you want to constrain n to be an integer, and if it is *not* an integer, the boundary condition won't be satisfied.)

(c) Attempt to repeat the process with $u_0(x) = x(1 - x)$. You should find that *Mathematica* solves the ODE with no difficulty, but cannot compute the inverse Laplace transform to recover u. (This is symptomatic of the fact that there is no formula for u in terms of elementary functions in this case.)

(d) To deal with the problem in (c), there is another method. Attempt to write $u_0(x) = x(1 - x)$ as a linear combination of $\sin(n\pi x)$ for different values of n. This gives what is called the *Fourier expansion* of the function. Since you know the solution to (E.6) and (E.7) in this case, and since the equation is linear, that enables you to write down the solution to the problem as an infinite series, using what is sometimes called the *principle of superposition*. To find the coefficients in the Fourier expansion, suppose you have a formal expansion

$$u_0(x) = x(1 - x) = \sum_{n=1}^{\infty} c_n \sin(n\pi x), \qquad 0 \le x \le 1. \qquad \text{(E.9)}$$

The coefficients c_n are computed as follows. Multiply both sides of (E.9) by $\sin(m\pi x)$ for fixed m, and integrate from 0 to 1. You can check using *Mathematica* that

$$\int_0^1 \sin^2(n\pi x)\, dx = \frac{1}{2}, \qquad \int_0^1 \sin(n\pi x)\sin(m\pi x)\, dx = 0, \qquad n \ne m.$$

Thus one gets

$$\int_0^1 \left(\sum_{n=1}^{\infty} c_n \sin(n\pi x)\right) \sin(m\pi x)\, dx$$

$$= \sum_{n=1}^{\infty} c_n \int_0^1 \sin(n\pi x)\sin(m\pi x)\, dx$$

$$= \sum_{n=1}^{\infty} c_n \left(\tfrac{1}{2} \text{ if } m = n, 0 \text{ otherwise}\right) = \frac{c_m}{2},$$

and so

$$c_m = 2\int_0^1 x(1-x)\sin(m\pi x)\, dx.$$

Compute the coefficients c_m for $1 \leq m \leq 10$ and plot

$$u_0(x) \quad \text{and} \quad \sum_{n=1}^{10} c_n \sin(n\pi x)$$

on the same axes to see the agreement. Then find the corresponding solution of (E.6) and (E.7) and plot it as a function of x and t. (The plot will be a surface in three dimensions. To plot the surface with *Mathematica* use the command **Plot3D**, which uses syntax similar to **Plot**. It should suffice to take $0 \leq x, t \leq 1$.) How does the solution match your intuition that temperature fluctuations in the rod at $t = 0$ should dissipate with time?

19. Consider the mass-damper system with a dead zone as depicted in Figure E.1. The motion of the system is described by the following combination of equations:

$$\begin{cases} my'' + 2cy' + k(y+b) = F(t), & y < -b \\ my'' + 2cy' = F(t), & -b \leq y \leq b \\ my'' + 2cy' + k(y-b) = F(t), & y > b. \end{cases} \qquad \text{(E.10)}$$

(a) Rewrite equation (E.10) as a single equation using the **HeavisideTheta** function.

(b) Consider a free system, $F \equiv 0$, and let $m = 10\,\text{kg}$, $c = 50\,\text{kg/s}$ and $k = 150\,\text{kg/s}^2$. Suppose the motion starts at $y(0) = 0$ with initial velocity $v_0 = y'(0) = 20\,\text{m/s}$. Finally take $b = 0.5\,\text{m}$. Find a numerical solution with **NDSolve** and graph it on the interval $0 \leq t \leq 15$.

(c) Now by varying the friction constant c, the initial velocity v_0, and the dead zone size b, exhibit at least two additional types of behavior for the solution function.

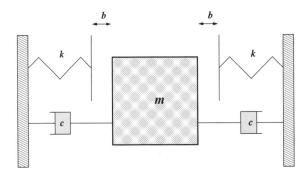

Figure E.1: A Mass-Damper System with a Dead Zone

(d) Repeat parts (b) and (c) in the presence of a forcing function

$$F(t) = 20\cos(12t).$$

20. Consider the "tunable" RLC circuit illustrated in Figure E.2. It consists of a variable resistor with resistance R, an inductance L, and a capacitor with capacitance C. The circuit can be tuned by adjusting R. If an input voltage $f(t)$ is applied as shown, the voltage v across the resistor satisfies the differential equation

$$v'' + \frac{R}{L}v' + \frac{1}{LC}v = \frac{R}{L}f'. \qquad (E.11)$$

Assume that in suitable units, $LC = 1$.

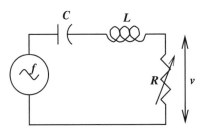

Figure E.2: An RLC Circuit with an Input Pulse

Solve the equation (E.11), with $r = \frac{R}{L}$ declared to be a symbolic variable and initial conditions $v(0) = v'(0) = 0$, using **NDSolve**, in the case where the input voltage $f(t) = u_1(t)$ is a step function. (Note that this means that $f'(t)$, which is what actually appears in the equation, is the Dirac delta function $\delta_1(t)$.) Plot the solution for $r = 0.1, 1, 3, 5, 10$.

Chapter 12

Higher Order Equations and Systems of First Order Equations

First and second order differential equations arise naturally in many applications. For example, first order differential equations occur in models of population growth and radioactive decay, and second order equations in the study of the motion of a falling body or the motion of a pendulum. There are several techniques for solving special classes of first and second order equations. These techniques produce symbolic solutions, expressed by a formula. For equations that cannot be solved by any of these techniques, we use numerical, geometric, or qualitative methods to investigate solutions.

Equations of higher order, and systems of first order equations, also arise naturally. For example, third order equations come up in fluid dynamics; fourth order equations in elasticity; and systems of first order equations in the study of spring–mass systems with several springs and masses. For general higher order equations and systems, there are only very few techniques for obtaining explicit formula solutions, and numerical, geometric, or qualitative techniques must be used. Nevertheless, for the special classes of constant coefficient linear equations and first order linear systems, there are techniques for producing explicit solutions.

In this chapter, we consider higher order equations and first order systems. We outline the basic theory in the linear case. We show how to solve linear first order systems, first by using *Mathematica* to compute eigenpairs of the coefficient matrix and then by using **DSolve**. Then we discuss the plotting of phase portraits, both using exact solutions and, as is generally necessary for nonlinear systems, using numerical solutions. In particular, at the end of this chapter we show how to use **NDSolve** to solve systems of equations numerically.

12.1   Higher Order Linear Equations

Consider the linear equation

$$y^{(n)} + p_1(t)y^{(n-1)} + \cdots + p_n(t)y = g(t), \qquad (12.1)$$

and assume that the coefficient functions are continuous on the interval $a < t < b$. Then the following are true:

- If $y_1(t), y_2(t), \ldots, y_n(t)$ are n linearly independent solutions of the corresponding homogeneous equation,

$$y^{(n)} + p_1(t)y^{(n-1)} + \cdots + p_n(t)y = 0, \qquad (12.2)$$

and if y_p is any particular solution of the inhomogeneous equation (12.1), then the general solutions of the equations (12.1), (12.2) are, respectively,

$$y(t) = \sum_{i=1}^{n} C_i y_i(t) + y_p(t)$$

and

$$y(t) = \sum_{i=1}^{n} C_i y_i(t).$$

- Given a point $t_0 \in (a, b)$ and initial conditions $y(t_0) = y_0,\ y'(t_0) = y_0',\ \ldots,$ $y^{(n-1)}(t_0) = y_0^{(n-1)}$, there is a unique solution to equation (12.1) satisfying these conditions; the solution has n continuous derivatives and exists on the entire interval (a, b).

- A set of n solutions $y_1(t), y_2(t), \ldots, y_n(t)$ of (12.2) is linearly independent if and only if the Wronskian

$$W(y_1, y_2, \ldots, y_n) = \det \begin{pmatrix} y_1 & y_2 & \cdots & y_n \\ y_1' & y_2' & \cdots & y_n' \\ \vdots & \vdots & \ddots & \vdots \\ y_1^{(n-1)} & y_2^{(n-1)} & \cdots & y_n^{(n-1)} \end{pmatrix}$$

is nonzero at some point in (a, b); a linearly independent collection $y_1, y_2, \ldots, y_n$ is called a *fundamental set* of solutions of (12.2).

We see that the general theory of higher order linear equations parallels that of first and second order linear equations. We note, however, that solutions may be difficult to find. In practice, we can find fundamental sets of solutions only in special cases such as constant coefficient and Euler equations. In the latter cases, solutions can be found by considering trial solutions e^{rt} and t^r, as in the second order case. There is a difficulty even for these equations if n is large, since it is difficult to compute the roots of the characteristic polynomial. We can, however, use *Mathematica* to find good approximations to the roots.

The command **DSolve** can solve many homogeneous constant coefficient and Euler equations. You can confirm this by solving several linear equations of higher order from your text.

12.2 Systems of First Order Equations

Consider the general system of first order equations,

$$
\begin{aligned}
x_1' &= F_1(t, x_1, x_2, \ldots, x_n) \\
x_2' &= F_2(t, x_1, x_2, \ldots, x_n) \\
&\vdots \\
x_n' &= F_n(t, x_1, x_2, \ldots, x_n).
\end{aligned}
\tag{12.3}
$$

Such systems arise in the study of spring–mass systems with many springs and masses, and in many other applications. Systems of first order equations also arise in the study of higher order equations. A single higher order differential equation

$$ y^{(n)} = F(t, y, y', \ldots, y^{(n-1)}) $$

can be converted into a system of first order equations by setting

$$ x_1 = y, x_2 = y', \ldots, x_n = y^{(n-1)}. $$

The resulting system is

$$
\begin{aligned}
x_1' &= x_2 \\
x_2' &= x_3 \\
&\vdots \\
x_{n-1}' &= x_n \\
x_n' &= F(t, x_1, x_2, \ldots, x_n).
\end{aligned}
$$

It is not surprising that the problem of analyzing the general system (12.3) is quite formidable. Generally, the techniques we can bring to bear will be numerical, geometric, or qualitative. We shall discuss the application of these techniques to nonlinear equations at the end of this chapter and in Chapter 13. For now, we merely observe that we can expect to make some progress toward a formula solution for *linear systems*.

12.2.1   Linear First Order Systems

Consider a linear system

$$
\begin{aligned}
x_1' &= a_{11}(t)x_1 + \cdots + a_{1n}(t)x_n + b_1(t) \\
x_2' &= a_{21}(t)x_1 + \cdots + a_{2n}(t)x_n + b_2(t) \\
&\;\;\vdots \\
x_n' &= a_{n1}(t)x_1 + \cdots + a_{nn}(t)x_n + b_n(t).
\end{aligned}
$$

We can write this system in matrix notation:

$$
\mathbf{X}' = \mathbf{A}(t)\mathbf{X} + \mathbf{B}(t),
$$

where

$$
\mathbf{X}(t) = \begin{pmatrix} x_1(t) \\ x_2(t) \\ \vdots \\ x_n(t) \end{pmatrix}, \qquad
\mathbf{B}(t) = \begin{pmatrix} b_1(t) \\ b_2(t) \\ \vdots \\ b_n(t) \end{pmatrix},
$$

and

$$
\mathbf{A}(t) = \begin{pmatrix} a_{11}(t) & a_{12}(t) & \cdots & a_{1n}(t) \\ \vdots & \vdots & \ddots & \vdots \\ a_{n1}(t) & a_{n2}(t) & \cdots & a_{nn}(t) \end{pmatrix}.
$$

There is a general theory for linear systems, which is completely parallel to the theory for a single linear equation. If $\mathbf{X}^{(1)}, \mathbf{X}^{(2)}, \ldots, \mathbf{X}^{(n)}$ is a family of linearly independent solutions, *i.e.*, a *fundamental set* of solutions, of the homogeneous equation (with $\mathbf{B} = \mathbf{0}$), and if $\mathbf{X}_p$ is a particular solution of the inhomogeneous equation, then the general solutions of these equations are, respectively,

$$
\mathbf{X} = \sum_{i=1}^{n} C_i \mathbf{X}^{(i)}
$$

and

$$
\mathbf{X} = \sum_{i=1}^{n} C_i \mathbf{X}^{(i)} + \mathbf{X}_p.
$$

The test for linear independence of n solutions is

$$
\det(\mathbf{X}^{(1)} \ldots \mathbf{X}^{(n)}) \neq 0.
$$

For linear systems, the constant coefficient case is the easiest to handle, just as it was for single linear equations. Consider the homogeneous linear system

$$
\mathbf{X}' = \mathbf{A}\mathbf{X}, \tag{12.4}
$$

where

$$\mathbf{A} = \begin{pmatrix} a_{11} & a_{12} & \cdots & a_{1n} \\ \vdots & \vdots & \ddots & \vdots \\ a_{n1} & a_{n2} & \cdots & a_{nn} \end{pmatrix}$$

is an $n \times n$ matrix of real constants. We seek nonzero solutions of the form $\mathbf{X}(t) = \boldsymbol{\xi} e^{rt}$; note the similarity to the case of a single linear equation. We find that $\mathbf{X}(t)$ is a solution of (12.4) if and only if

$$\mathbf{A}\boldsymbol{\xi} = r\boldsymbol{\xi},$$

i.e., if and only if r is an *eigenvalue* of $\mathbf{A}$ and $\boldsymbol{\xi}$ is a corresponding *eigenvector*. We call the pair $r, \boldsymbol{\xi}$ an *eigenpair*. In order to explain how to construct a fundamental set of solutions, we have to consider three separate cases.

Case I. A has distinct real eigenvalues.

Let $r_1, r_2, \ldots, r_n$ be the distinct eigenvalues of $\mathbf{A}$, and let $\boldsymbol{\xi}^{(1)}, \boldsymbol{\xi}^{(2)}, \ldots, \boldsymbol{\xi}^{(n)}$ be corresponding eigenvectors. Then

$$\mathbf{X}^{(1)}(t) = \boldsymbol{\xi}^{(1)} e^{r_1 t}, \ldots, \mathbf{X}^{(n)}(t) = \boldsymbol{\xi}^{(n)} e^{r_n t}$$

is a fundamental set of solutions for (12.4), and

$$\mathbf{X}(t) = c_1 \boldsymbol{\xi}^{(1)} e^{r_1 t} + \cdots + c_n \boldsymbol{\xi}^{(n)} e^{r_n t}$$

is the general solution.

Case II. A has distinct eigenvalues, some of which are complex.

Since $\mathbf{A}$ is real, if $r, \boldsymbol{\xi}$ is a complex eigenpair, then the complex conjugate $\bar{r}, \bar{\boldsymbol{\xi}}$ is also an eigenpair. Thus the corresponding solutions

$$\mathbf{X}^{(1)}(t) = \boldsymbol{\xi} e^{rt}, \quad \mathbf{X}^{(2)}(t) = \bar{\boldsymbol{\xi}} e^{\bar{r}t}$$

are conjugate. Therefore, we can find two real solutions of (12.4) corresponding to the pair $r, \bar{r}$ by taking the real and imaginary parts of $\mathbf{X}^{(1)}(t)$ or $\mathbf{X}^{(2)}(t)$. Writing $\boldsymbol{\xi} = \mathbf{a} + i\mathbf{b}$, where $\mathbf{a}$ and $\mathbf{b}$ are real, and $r = \lambda + i\mu$, where λ and μ are real, we have

$$\begin{aligned} \mathbf{X}^{(1)}(t) &= (\mathbf{a} + i\mathbf{b})e^{(\lambda + i\mu)t} \\ &= (\mathbf{a} + i\mathbf{b})e^{\lambda t}(\cos \mu t + i \sin \mu t) \\ &= e^{\lambda t}(\mathbf{a} \cos \mu t - \mathbf{b} \sin \mu t) + i e^{\lambda t}(\mathbf{a} \sin \mu t + \mathbf{b} \cos \mu t). \end{aligned}$$

Thus the vector functions

$$\mathbf{u}(t) = e^{\lambda t}(\mathbf{a} \cos \mu t - \mathbf{b} \sin \mu t)$$

$$\mathbf{v}(t) = e^{\lambda t}(\mathbf{a} \sin \mu t + \mathbf{b} \cos \mu t)$$

are real solutions to (12.4).

To keep the discussion simple, assume that $r_1 = \lambda + i\mu$, $r_2 = \lambda - i\mu$ are complex, and that $r_3, \ldots, r_n$ are real and distinct. Let the corresponding eigenvectors be $\boldsymbol{\xi}^{(1)} = \mathbf{a} + i\mathbf{b}, \boldsymbol{\xi}^{(2)} = \mathbf{a} - i\mathbf{b}, \boldsymbol{\xi}^{(3)}, \ldots, \boldsymbol{\xi}^{(n)}$. Then

$$\mathbf{u}(t), \mathbf{v}(t), \boldsymbol{\xi}^{(3)} e^{r_3 t}, \ldots, \boldsymbol{\xi}^{(n)} e^{r_n t}$$

is a fundamental set of solutions for (12.4). The general situation should now be clear.

Case III. A has repeated eigenvalues.

We restrict our discussion to the case $n = 2$. Suppose $r = \rho$ is an eigenvalue of $\mathbf{A}$ of multiplicity 2, meaning that ρ is a double root of the characteristic polynomial $\det(\mathbf{A} - r\mathbf{I})$. There are still two possibilities. If we can find two linearly independent eigenvectors $\boldsymbol{\xi}^{(1)}$ and $\boldsymbol{\xi}^{(2)}$ corresponding to ρ, then the solutions $\mathbf{X}^{(1)}(t) = \boldsymbol{\xi}^{(1)} e^{\rho t}$ and $\mathbf{X}^{(2)}(t) = \boldsymbol{\xi}^{(2)} e^{\rho t}$ form a fundamental set. The other possibility is that there is only one (linearly independent) eigenvector $\boldsymbol{\xi}$ with eigenvalue ρ. Then one solution of (12.4) is given by

$$\mathbf{X}^{(1)}(t) = \boldsymbol{\xi} e^{\rho t},$$

and a second by

$$\mathbf{X}^{(2)}(t) = \boldsymbol{\xi} t e^{\rho t} + \boldsymbol{\eta} e^{\rho t},$$

where $\boldsymbol{\eta}$ satisfies

$$(\mathbf{A} - \rho\mathbf{I})\boldsymbol{\eta} = \boldsymbol{\xi}.$$

The vector $\boldsymbol{\eta}$ is called a *generalized eigenvector* of $\mathbf{A}$ for the eigenvalue ρ.

12.2.2 Using *Mathematica* to Find Eigenpairs

We now show how to use *Mathematica* to find the eigenpairs of a matrix, and thus to find a fundamental set of solutions and the general solution of a system of linear differential equations with constant coefficients.

Example 12.1 Consider the system

$$\mathbf{x}' = \begin{pmatrix} 3 & -2 & 0 \\ 2 & -2 & 0 \\ 0 & 1 & 1 \end{pmatrix} \mathbf{x}.$$

We enter the coefficient matrix in *Mathematica* as a list of lists:

```
A = {{3, -2, 0}, {2, -2, 0}, {0, 1, 1}}
```

```
{{3, -2, 0}, {2, -2, 0}, {0, 1, 1}}
```

We can display A in standard matrix form using the command:

```
MatrixForm[A]
```

$$\begin{pmatrix} 3 & -2 & 0 \\ 2 & -2 & 0 \\ 0 & 1 & 1 \end{pmatrix}$$

We find the eigenpairs of $\mathbf{A}$ with the command:

```
eigsys = Eigensystem[A]
```

$$\{\{2, -1, 1\}, \{\{2, 1, 1\}, \{-1, -2, 1\}, \{0, 0, 1\}\}\}$$

The output of **Eigensystem** is a list containing first a list of the eigenvalues and then a list of the corresponding eigenvectors. In this example, the eigenpairs are:

$$r_1 = 2, \ \boldsymbol{\xi}^{(1)} = \begin{pmatrix} 2 \\ 1 \\ 1 \end{pmatrix},$$

$$r_2 = -1, \ \boldsymbol{\xi}^{(2)} = \begin{pmatrix} -1 \\ -2 \\ 1 \end{pmatrix},$$

$$r_3 = 1, \ \boldsymbol{\xi}^{(3)} = \begin{pmatrix} 0 \\ 0 \\ 1 \end{pmatrix}.$$

Thus a fundamental set of solutions is

$$\mathbf{X}^{(1)}(t) = \begin{pmatrix} 2 \\ 1 \\ 1 \end{pmatrix} e^{2t}, \quad \mathbf{X}^{(2)}(t) = \begin{pmatrix} -1 \\ -2 \\ 1 \end{pmatrix} e^{-t}, \quad \mathbf{X}^{(3)}(t) = \begin{pmatrix} 0 \\ 0 \\ 1 \end{pmatrix} e^{t},$$

and the general solution is

$$\mathbf{x}(t) = c_1 \begin{pmatrix} 2 \\ 1 \\ 1 \end{pmatrix} e^{2t} + c_2 \begin{pmatrix} -1 \\ -2 \\ 1 \end{pmatrix} e^{-t} + c_3 \begin{pmatrix} 0 \\ 0 \\ 1 \end{pmatrix} e^{t}.$$

Suppose we are seeking the specific solution with initial value

$$\mathbf{x}(0) = \begin{pmatrix} 3 \\ 5 \\ 0 \end{pmatrix}.$$

Then the constants c_1, c_2, and c_3 must satisfy

$$c_1 \begin{pmatrix} 2 \\ 1 \\ 1 \end{pmatrix} + c_2 \begin{pmatrix} -1 \\ -2 \\ 1 \end{pmatrix} + c_3 \begin{pmatrix} 0 \\ 0 \\ 1 \end{pmatrix} = \begin{pmatrix} 2 & -1 & 0 \\ 1 & -2 & 0 \\ 1 & 1 & 1 \end{pmatrix} \begin{pmatrix} c_1 \\ c_2 \\ c_3 \end{pmatrix} = \begin{pmatrix} 3 \\ 5 \\ 0 \end{pmatrix}.$$

Note that the columns of the matrix in the above equation are the eigenvectors of the coefficient matrix $\mathbf{A}$. To solve this linear system, we use the command **LinearSolve**. Let $\mathbf{b}$ be the column vector on the right-hand side and $\mathbf{M}$ be the coefficient matrix:

```
b = {3, 5, 0};
M = Transpose[Last[eigsys]];
LinearSolve[M, b]
```

$$\left\{\frac{1}{3}, \frac{-7}{3}, 2\right\}$$

Thus the solution of the linear system is $c_1 = 1/3$, $c_2 = -7/3$, $c_3 = 2$, and the solution to our initial value problem is

$$\mathbf{x}(t) = \frac{1}{3}\begin{pmatrix} 2 \\ 1 \\ 1 \end{pmatrix} e^{2t} - \frac{7}{3}\begin{pmatrix} -1 \\ -2 \\ 1 \end{pmatrix} e^{-t} + 2\begin{pmatrix} 0 \\ 0 \\ 1 \end{pmatrix} e^{t}.$$

Example 12.2 Consider the system

$$\mathbf{x}' = \begin{pmatrix} 3 & -2 \\ 4 & -1 \end{pmatrix} \mathbf{x} \qquad \text{(Boyce and DiPrima, Sect. 7.6, Prob. 1).}$$

We solve the corresponding eigensystem as follows:

```
A = {{3, -2}, {4, -1}};
Eigensystem[A]
```

$$\left\{\{1 + 2i, 1 - 2i\}, \left\{\left\{\frac{1}{2} + \frac{i}{2}, 1\right\}, \left\{\frac{1}{2} - \frac{i}{2}, 1\right\}\right\}\right\}$$

So the eigenpairs are

$$1 + 2i, \begin{pmatrix} (1+i)/2 \\ 1 \end{pmatrix}$$

and

$$1 - 2i, \begin{pmatrix} (1-i)/2 \\ 1 \end{pmatrix}.$$

Note that the second pair is the complex conjugate of the first. Thus a complex fundamental set of solutions is

$$\mathbf{X}^{(1)}(t) = \begin{pmatrix} (1+i)/2 \\ 1 \end{pmatrix} e^{(1+2i)t}, \quad \mathbf{X}^{(2)}(t) = \begin{pmatrix} (1-i)/2 \\ 1 \end{pmatrix} e^{(1-2i)t}.$$

The real and imaginary parts $\mathbf{u}(t)$ and $\mathbf{v}(t)$ of the solution $\mathbf{X}^{(1)}(t)$ can be extracted by hand or by using the *Mathematica* commands **ComplexExpand**, **Re**, and **Im** as follows:

```
ComplexExpand[Re[{1/2 + i/2, 1} Exp[(1 + 2i)t]]]
ComplexExpand[Im[{1/2 + i/2, 1} Exp[(1 + 2i)t]]]
```

$$\left\{ \frac{1}{2} e^t \operatorname{Cos}[2t] \ - \ \frac{1}{2} e^t \operatorname{Sin}[2t], \ e^t \operatorname{Cos}[2t] \right\}$$

$$\left\{ \frac{1}{2} e^t \operatorname{Cos}[2t] \ + \ \frac{1}{2} e^t \operatorname{Sin}[2t], \ e^t \operatorname{Sin}[2t] \right\}$$

Either way, we see that a real fundamental set of solutions is

$$\mathbf{u}(t) = \frac{1}{2} e^t \left[\begin{pmatrix} 1 \\ 2 \end{pmatrix} \cos 2t + \begin{pmatrix} -1 \\ 0 \end{pmatrix} \sin 2t \right]$$

$$\mathbf{v}(t) = \frac{1}{2} e^t \left[\begin{pmatrix} 1 \\ 0 \end{pmatrix} \cos 2t + \begin{pmatrix} 1 \\ 2 \end{pmatrix} \sin 2t \right].$$

Example 12.3 Consider the system

$$\mathbf{x}' = \begin{pmatrix} 3 & -4 \\ 1 & -1 \end{pmatrix} \mathbf{x} \qquad \text{(Boyce and DiPrima, Sect. 7.8, Prob. 1)}.$$

We solve the corresponding eigensystem as follows:

```
A = {{3, -4}, {1, -1}};
Eigensystem[A]
```

```
{{1, 1}, {{2, 1}, {0, 0}}}
```

The number 1 is listed twice as an eigenvalue, and $\{0, 0\}$ is listed as an eigenvector (although it really isn't one because eigenvectors must be nonzero). This is *Mathematica*'s way of reporting that 1 is an eigenvalue of multiplicity 2 and that $\begin{pmatrix} 2 \\ 1 \end{pmatrix}$ is the only corresponding eigenvector. Let $\rho = 1$ and $\boldsymbol{\xi} = \begin{pmatrix} 2 \\ 1 \end{pmatrix}$. Then

$$\mathbf{X}^{(1)}(t) = \boldsymbol{\xi} e^{\rho t} = \begin{pmatrix} 2 \\ 1 \end{pmatrix} e^t$$

is one solution. To find a second solution we solve

$$(\mathbf{A} - \rho \mathbf{I})\boldsymbol{\eta} = \begin{pmatrix} 2 & -4 \\ 1 & -2 \end{pmatrix} \boldsymbol{\eta} = \boldsymbol{\xi} = \begin{pmatrix} 2 \\ 1 \end{pmatrix}.$$

We enter the coefficient matrix as **m** and let **b** = {2, 1}. We solve the corresponding linear system for η:

```
b = {2, 1};
M = {{2, -4}, {1, -2}};
η = LinearSolve[M, b]
```

```
{1, 0}
```

(This system actually has an infinite number of solutions; *Mathematica* returns one of them.) Thus a second solution of our system is

$$\mathbf{X}^{(2)}(t) = \boldsymbol{\xi}te^{\rho t} + \boldsymbol{\eta}e^{\rho t} = \begin{pmatrix} 2 \\ 1 \end{pmatrix} te^t + \begin{pmatrix} 1 \\ 0 \end{pmatrix} e^t.$$

The general solution is

$$\mathbf{X}(t) = c_1 \begin{pmatrix} 2 \\ 1 \end{pmatrix} e^t + c_2 \left[\begin{pmatrix} 2 \\ 1 \end{pmatrix} te^t + \begin{pmatrix} 1 \\ 0 \end{pmatrix} e^t \right].$$

We have shown how to use *Mathematica* to find eigenpairs, and thus solutions, for linear systems of differential equations with constant coefficients. The solutions to systems of linear equations can also be found with a direct application of **DSolve**.

Example 12.4 Consider the system $x' = y$, $y' = -x$. To find its general solution, type:

```
DSolve[{x'[t] == y[t], y'[t] == -x[t]}, {x[t], y[t]}, t]
```

```
{{x[t] → C[1] Cos[t] + C[2] Sin[t], y[t] → C[2] Cos[t] -
C[1] Sin[t]}}
```

To solve the same system with initial conditions $x(0) = 1$, $y(0) = 0$, you can type:

```
DSolve[{x'[t] == y[t], y'[t] == -x[t], x[0] == 1,
    y[0] == 0}, {x[t], y[t]}, t]
```

```
{{x[t] → Cos[t], y[t] → -Sin[t]}}
```

Both methods for solving linear systems are useful. Solving in terms of the eigenpairs of the coefficient matrix is somewhat involved, but yields simple and useful formulas for the solutions. The eigenpairs contain valuable information about the solutions, which we will exploit further in Problem Set F. On the other hand, it is simpler to use **DSolve**, and for many purposes the solutions generated by **DSolve** are completely satisfactory.

12.3   Phase Portraits

A solution of a 2×2 linear system is a pair of functions $x(t)$, $y(t)$. The plot of a solution as a function of t would be a curve in (t, x, y)-space. Generally, such three-dimensional plots are too complicated to be illuminating. Therefore, we project the curve from three-dimensional space into the (x, y)-plane. This plane is called the *phase plane*, and the resulting curve in the phase plane is called a *trajectory*. A trajectory is an example of a *parametric curve* and is drawn by plotting $(x(t), y(t))$ as the parameter t varies. A plot of a family of trajectories is called a *phase portrait* of the system.

In this section, we describe how to use *Mathematica*: (1) to solve an initial value problem consisting of a linear system and an initial condition, and plot the corresponding trajectory; (2) to solve a linear system using **DSolve** and draw a phase portrait; and (3) to solve a system using **NDSolve** and draw a phase portrait.

12.3.1   Plotting a Single Trajectory

Suppose we want to plot the trajectory of the initial value problem

$$x' = -3x + 2y, \quad y' = -x; \qquad x(0) = 1, \quad y(0) = 0.$$

First we enter the initial value problem, solve it using **DSolve**, and define a *Mathematica* function **solivp** using the formulas given by **DSolve**.

```
ivp = {x'[t] == -3 x[t] + 2 y[t], y'[t] == -x[t],
    x[0] == 1, y[0] == 0};
solivp[t_] = {x[t], y[t]} /. First[DSolve[ivp,
    {x[t], y[t]}, t]];
```

The function **solivp** can be evaluated at a particular time t, or at a sequence of times. To plot the trajectory on a given interval, say $-0.3 \le t \le 5$, we type the following:

```
ParametricPlot[solivp[t], {t, -0.3, 5}, PlotRange → All]
```

The result is shown in Figure 12.1. To use the sequence of commands above for plotting the solution of a different initial value problem, all you have to do is modify the definition of **ivp** and the range of **t** appropriately.

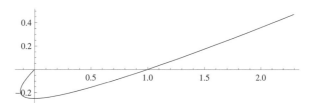

Figure 12.1: A Single Trajectory

12.3.2   Plotting Several Trajectories

Consider now the linear system

$$x' = x - 2y, \quad y' = -x.$$

We want to plot trajectories of this system with various initial conditions $x(0) = a$, $y(0) = b$. We first define a function **sol** that solves this system with initial conditions $x(0) = a$, $y(0) = b$. We first clear the definition of **b** from earlier in this chapter.

```
Clear[b]
ivp = {x'[t] == x[t] - 2y[t], y'[t] == -x[t], x[0] == a,
    y[0] == b};
```

```
sol[t_, a_, b_] = {x[t], y[t]} /. First[DSolve[ivp,
    {x[t], y[t]}, t]]
```

The function **sol** can be evaluated for a particular choice of initial conditions $x(0) = a$, $y(0) = b$, at a particular value of t.

Next we plot a phase portrait of the linear system. We do this by creating a table of solutions with different initial conditions, and then using **ParametricPlot** to plot this table of solution functions.

```
ParametricPlot[Evaluate[Flatten[Table[sol[t, a, b],
    {a, -2, 2}, {b, -2, 2}], 1]], {t, -3, 3},
    PlotRange → {-10, 10}]
```

The **Table** command embedded in this definition produces a list with two levels of braces; we used **Flatten[...,1]** to remove the unnecessary braces. (Though in this example the extra braces don't cause a problem, in some cases they can prevent the plotting command from working. Our use of **Evaluate** here also is not strictly necessary, but we include it for the reasons explained in Section 8.10.) Here we have chosen a rectangular grid of 25 initial conditions, with a ranging from -2 to 2 and b ranging from -2 to 2 (in integer increments). We also chose a time interval of -3 to 3, and a range of -10 to 10 for the x and y axes. The resulting phase portrait is shown in Figure 12.2. The time interval must be chosen somewhat carefully; a small time interval will result in solution curves that are too short to get an idea of where they're headed, while an unnecessarily large time interval can cause computational problems.

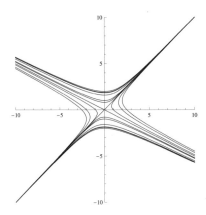

Figure 12.2: A Family of Curves

You should be able to plot the phase portraits for any 2×2 linear system with constant coefficients using the procedure above, replacing the differential equations in the definition of **ivp**, and choosing, by trial and error, an appropriate time interval and appropriate initial conditions to use in defining the ranges for **t**, **a**, and **b**. By combining the phase portrait with a plot of the vector field (see Chapter 13), you should be able to determine the direction of increasing t on the curves. Also, in each case you should be able to decide whether the origin is a *sink* (all solutions approach the origin as t increases), *source* (all solutions move away from the origin), *saddle point* (solutions approach the origin along one direction, then move away from the origin in another direction), or *center* (solutions follow closed curves around the origin).

The procedure for plotting a phase portrait must be modified slightly if you wish to vary only one parameter in the initial data. For example, consider the initial value problem

$$x' = x + 2y, \quad y' = -x; \quad x(0) = a, \quad y(0) = 0$$

for various values of a. The commands

```
ivp = {x'[t] == x[t] + 2 y[t], y'[t] == -x[t], x[0] == a,
    y[0] == 0};
sol[t_, a_] = {x[t], y[t]} /. First[DSolve[ivp,
    {x[t], y[t]}, t]]
ParametricPlot[Evaluate[Table[sol[t, a], {a, -4, 4}]],
    {t, -10, 10}, PlotRange → {-10, 10}]
```

yields the nine solutions shown in Figure 12.3. (Count the curves! One of the solutions is the equilibrium solution $x(t) = 0$, $y(t) = 0$.)

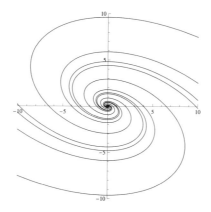

Figure 12.3: Another Family of Curves

12.3.3   Numerical Solutions of First Order Systems

The symbolic solver **DSolve** can solve any homogeneous 2×2 linear system with constant coefficients. For inhomogeneous linear systems, linear systems with variable coefficients, and nonlinear systems, **DSolve** will generally not do the job. In these situations, we turn to the numerical solver **NDSolve**. Note that we cannot simply replace **DSolve** with **NDSolve** in the commands above, in part because the syntax of **NDSolve** is different from that of **DSolve**, but more importantly, because **NDSolve** must have numerical values for the initial conditions and the time interval before it is evaluated.

Suppose we wish to solve the initial value problem

$$\begin{aligned} x_1' &= F_1(t, x_1, x_2, \ldots, x_n) & x_1(t_0) &= x_{1,0} \\ x_2' &= F_2(t, x_1, x_2, \ldots, x_n) & x_2(t_0) &= x_{2,0} \\ &\;\;\vdots & &\;\;\vdots \\ x_n' &= F_n(t, x_1, x_2, \ldots, x_n) & x_n(t_0) &= x_{n,0} \end{aligned}$$

for a first order system of differential equations. We can use **NDSolve** on this system much as we used it on a single first order equation. We illustrate with the initial value problem

$$x' = x - 2y, \quad y' = -x; \qquad x(0) = 2, \quad y(0) = -2$$

on the interval $0 \le t \le 2$. First we make a table of approximate solution values for t in increments of 0.2.

```
ivp = {x'[t] == x[t] - 2y[t], y'[t] == -x[t], x[0] == 2,
    y[0] == -2};
nsolivp[t_] = {x[t], y[t]} /. First[NDSolve[ivp,
    {x[t], y[t]}, {t, -3, 3}]];
TableForm[Table[Join[{t}, nsolivp[t]], {t, 0, 2, 0.2}],
    TableHeadings -> {None, {"t", "x(t)", "y(t)"}},
    TableSpacing -> {1, 5}]
```

t	x[t]	y[t]
0.	2.	-2.
0.2	3.43238	-2.53492
0.4	5.48789	-3.41427
0.6	8.48777	-4.7927
0.8	12.9085	-6.90359
1.	19.4589	-10.0973
1.2	29.1943	-14.8984
1.4	43.688	-22.0906
1.6	65.2854	-32.8446
1.8	97.4851	-48.9078
2.	145.505	-72.8877

You can also plot the numerical solution **nsolivp** using **ParametricPlot**, for example.

To plot a phase portrait for the system

$$x' = x - 2y, \quad y' = -x$$

using **NDSolve**, we use a slightly different approach than before. First, we define a function that inputs values of a and b and then numerically solves the system with initial conditions $x(0) = a, y(0) = b$ for t between -3 and 3:

```
nsol[a_, b_] := {x[t], y[t]} /. First[NDSolve[{x'[t] ==
    x[t] - 2y[t], y'[t] == -x[t], x[0] == a, y[0] == b},
    {x[t], y[t]}, {t, -3, 3}]]
```

Here we used delayed evaluation (**:=**) to prevent *Mathematica* from trying to evaluate **NDSolve** until we supply numerical values for **a** and **b**. Also, we do not input **t** into **nsol** because we must wait until **NDSolve** is evaluated before substituting a numerical value for **t**. To obtain a numerical solution at $t = 1$ for initial conditions $x(0) = 2, y(0) = -2$, you can type **nsol[2, -2] /. t → 1**. To produce a phase portrait similar to Figure 12.2, type

```
ParametricPlot[Evaluate[Flatten[Table[nsol[a, b],
    {a, -2, 2}, {b, -2, 2}], 1]], {t, -3, 3},
    PlotRange → {-10, 10}]
```

Note that to enlarge the time interval for the phase portrait, you must do so in the definition of **nsol** as well as in the **ParametricPlot** command.

Though we have used linear systems as examples in this section, numerical methods are most useful for nonlinear systems, for which solutions generally cannot be found by symbolic methods. The syntax for using **NDSolve** on nonlinear systems is exactly the same as in the examples above. For example, consider the nonlinear system

$$x' = x\left(1 - x^2 - y^2\right) + \frac{x^2}{\sqrt{x^2 + y^2}} - y, \qquad y' = y\left(1 - x^2 - y^2\right) + \frac{xy}{\sqrt{x^2 + y^2}} + x.$$

The following commands generate the phase portrait shown in Figure 12.4:

```
nsol[a_,b_] := {x[t], y[t]} /. First[NDSolve[
    {x'[t] == x[t] (1 - x[t]^2 - y[t]^2) + x[t]^2/Sqrt[x[t]^2 + y[t]^2] - y[t],
    y'[t] == y[t] (1 - x[t]^2 - y[t]^2) + x[t] y[t]/Sqrt[x[t]^2 + y[t]^2] + x[t],
    x[0] == a, y[0] == b}, {x[t], y[t]}, {t, -3, 3}]]
ParametricPlot[Evaluate[Flatten[Table[nsol[a, b],
    {a, -2, 2}, {b, -2, 2}], 1]], {t, 0, 2},
    PlotRange → {-10, 10}]
```

Higher order differential equations are handled by converting them into a first order system, as explained in Section 12.2, and by applying **NDSolve** to the resulting system. In Section 9.1 we showed how to numerically solve a second order equation in this manner.

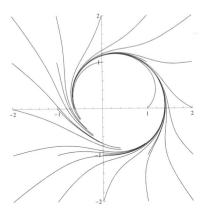

Figure 12.4: A Family of Curves for a Nonlinear System

Mathematica also has a graphical user interface called **EquationTrekker** that can plot a family of solutions for a system of two first order differential equations, or a single first or second order equation, using initial conditions selected with the mouse. After loading its package with **<<EquationTrekker`**, you can run **EquationTrekker** with the same syntax as **NDSolve**, except that you must omit the initial conditions. For example, to plot solutions of the system

$$x' = x - 2y, \quad y' = -x,$$

you can enter:

```
EquationTrekker[{x'[t] == x[t] - 2y[t], y'[t] == -x[t]},
    {x[t], y[t]}, {t, -3, 3}]
```

A new window will appear. To specify initial conditions at $t = 0$ (instead of $t = -3$), and thus get *Mathematica* to solve both forward and backward in time, change the value in the box next to t_0 from -3 to 0, and click the arrow next to the box. Then right-click in the box with the axes, and you should get a curve similar to those in Figure 12.2. You can then right-click on another initial condition, *etc.* See the online help for other capabilities. While you have the **EquationTrekker** window open, you will not be able to enter other *Mathematica* commands. When you close the window, your plot will be reproduced in your Notebook and you can continue your *Mathematica* session.

Chapter 13

Qualitative Theory for Systems of Differential Equations

In this chapter, we extend the qualitative theory of autonomous differential equations from single equations to systems of two equations. Thus we consider a system of the form

$$\begin{cases} x' = F(x, y) \\ y' = G(x, y). \end{cases} \tag{13.1}$$

We assume that the functions F and G have continuous partial derivatives everywhere in the plane, and that the critical points—the common zeros of F and G—are isolated.

We know from the fundamental existence and uniqueness theorems that, corresponding to each choice of initial data (x_0, y_0), there is a unique pair of functions $x(t)$, $y(t)$ satisfying the system (13.1) of differential equations and the initial conditions $x(0) = x_0$, $y(0) = y_0$. Taken together, this pair of functions parameterizes a curve, or trajectory, in the *phase plane* passing through the point (x_0, y_0). Since we are dealing with autonomous systems (*i.e.*, the variable t does not appear on the right side of (13.1)), the solution curves are *independent of the starting time*. That is, if we consider initial conditions $x(t_0) = x_0$, $y(t_0) = y_0$, then we get the same solution curve with a time delay of t_0 units. Moreover, we also know that two distinct trajectories cannot intersect. Thus the plane is covered by the family of trajectories; the plot of these curves is the *phase portrait* of (13.1).

As we know, we can find explicit formula solutions $x(t)$, $y(t)$ only for simple systems. Thus we are forced to turn to qualitative and numerical methods. It is our purpose here to discuss qualitative techniques for studying the solutions of (13.1).

In most cases of physical interest, every solution curve behaves in one of the following ways:

(i) The curve consists of a single critical point.

(ii) The curve tends to a critical point as $t \to \infty$.

(iii) The curve is unbounded: the distance from $(x(t), y(t))$ to the origin becomes arbitrarily large as $t \to \infty$, or as $t \to t^*$, for some finite t^*.

(iv) The curve is periodic: the parametric functions satisfy $x(t + t_p) = x(t)$, $y(t + t_p) = y(t)$ for some fixed t_p.

(v) The curve approaches a periodic solution, spiraling in on a circle for example.

Qualitative techniques may be able to identify the kinds of solutions that appear. In order to keep matters simple, we try to answer the following specific questions:

(a) What can we say about the critical points? Can we make a qualitative guess about their nature and stability? Can we deduce anything about the solution curves that start out close to the critical points?

(b) Can we predict anything about the long-term behavior of the solution curves? This includes those that come close to the critical points, as well as the solution curves in general.

The list of possible limiting behaviors of solution curves is based on the Poincaré-Bendixson Theorem (Theorem 9.7.3 in Boyce and DiPrima), which is valid for autonomous systems of two equations. Systems of three or more equations can have much more complicated limiting behavior. For example, solution curves can remain bounded and yet fail to approach any equilibrium or periodic state as $t \to \infty$. This phenomenon, called "chaos" by mathematicians, was anticipated by Poincaré (and perhaps even earlier by the physicist Maxwell), but most scientists did not appreciate how widespread chaos is until the arrival of computers.

There are two qualitative methods for analyzing systems of equations: one based on the idea of a *vector field*; and the other based on *linearized stability analysis*. The latter method is treated in detail in most differential equations texts; it provides information about the stability of critical points of a nonlinear system by studying associated linear systems. In this chapter, we focus on the former method: the use of vector fields in qualitative analysis. Some of the problems in Problem Set F involve linearized stability analysis; see the solution to Problem 6 in the *Sample Solutions* for an example that uses *Mathematica* for this analysis.

Here is the basic scenario for vector field analysis. For any solution curve $(x(t), y(t))$ and any point t, we have the differential equations

$$\begin{cases} x'(t) = F(x(t), y(t)) \\ y'(t) = G(x(t), y(t)). \end{cases}$$

If we employ vector notation

$$\mathbf{x}(t) = \begin{bmatrix} x(t) \\ y(t) \end{bmatrix}, \quad \mathbf{f}(\mathbf{x}) = \begin{bmatrix} F(\mathbf{x}) \\ G(\mathbf{x}) \end{bmatrix},$$

then the system is written as

$$\mathbf{x}' = \mathbf{f}(\mathbf{x}).$$

Moreover, the vector $\mathbf{x}'(t) = \mathbf{f}(\mathbf{x}(t))$ is just the tangent vector to the curve $\mathbf{x}(t)$. Since knowledge of the collection of tangent vectors to the solution curves gives a good idea of the curves themselves, it would be useful to have a plot of these vectors. Thus, corresponding to any vector $\mathbf{x}$, we draw the vector $\mathbf{f}(\mathbf{x})$, translated so that its foot is at the point $\mathbf{x}$. This plot is called the vector field of the system (13.1). Since we cannot do this at every point of the phase plane, we only draw the vectors at a set of regularly chosen points. Then we step back and look at the resulting vector field. We do not have the solution curves sketched, but we do have a representative collection of their tangent vectors. From the tangent vectors, we can get a good idea of the curves themselves. Indeed, we can often answer the questions above by carefully examining the vector field.

Vector fields can be drawn by hand for simple systems, but the *Mathematica* command **VectorFieldPlot** can draw vector fields of any first order system of two equations. The use of **VectorFieldPlot** to draw vector fields is similar to its use to draw direction fields of single first order equations (see Chapter 6). We illustrate the drawing of vector fields by examining two specific systems, namely

$$\begin{cases} x' = x(1 - x - y) \\ y' = y(0.75 - 0.5x - y) \end{cases} \tag{13.2}$$

and

$$\begin{cases} x' = x(5 - x - y) \\ y' = y(-2 + x). \end{cases} \tag{13.3}$$

System (13.2) is a competing species model, which is discussed in Boyce and DiPrima, Example 1, Section 9.4. System (13.3) is a predator-prey model with logistic growth (for the prey), which is discussed in Boyce and DiPrima, Problems 11 and 12, Section 9.5.

The following sequence of commands draws a direction field for system (13.2):

```
<< VectorFieldPlots`
VectorFieldPlot[{x (1 - x - y), y (0.75 - 0.5 x - y)},
    {x, 0, 1.5}, {y, 0, 1.5}, Frame → True,
    FrameLabel → {x, y}];
```

The result is shown in Figure 13.1. Note that by default *Mathematica* will plot 15 points (vectors) in each direction, resulting in a 15×15 grid. When drawing vector fields, you should generally plot 15 to 30 vectors in each direction. The number of vectors can be adjusted using the option **PlotPoints**.

In a vector field, the vector drawn at the point $\mathbf{x} = (x, y)$ indicates both the direction and length of $\mathbf{f}(\mathbf{x})$. The vector at $\mathbf{x}$ drawn by **VectorFieldPlot** faithfully represents the direction of $\mathbf{f}(\mathbf{x})$, but its length is only proportional to the length of $\mathbf{f}(\mathbf{x})$. The critical, or equilibrium, points are those points at which the vector field vanishes, *i.e.*, those points where $F(x, y) = G(x, y) = 0$. Thus the vectors are very short near the critical points. In Figure 13.1 it is difficult to tell exactly where the critical points are. Since most of the vectors are so short that we see only the arrowheads, we cannot tell which ones are really shortest. Nonetheless, we can approximately locate the critical points as places where the direction of the arrows changes sharply as their location changes.

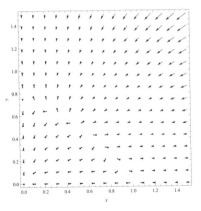

Figure 13.1: Vector Field for System (13.2)

We can improve the picture by changing the relative sizes of the vectors. One way to do this would be to make the vectors have a constant length. This can be done with the option `ScaleFunction → (1&)`. The `&` indicates to *Mathematica* that the constant "1" is to be interpreted as a pure function. A better approach is to use a scaling function that lengthens the vectors near the critical points enough that their directions are clearly indicated, but not so much that the relative lengths of the vectors are obscured. For example, we can use an option of the form `ScaleFunction → (# + c &)` for some constant c. The symbol `#` refers to the argument of the scaling function. Thus a vector of length proportional to r is replaced by a vector of length proportional to $r+c$. The constant c can be found by trial and error; the idea is to make the tails of the arrows visible, but short, near the critical points. Increasing c lengthens the shorter arrows, but a value of c that is too large will make all the arrows nearly the same length. Ideally, c should be reasonably close to the maximum value of the length of the vector field in the given plotting rectangle. We find that a value of $c = 5$ works well in this example:

```
VectorFieldPlot[{x (1 - x - y), y (0.75 - 0.5 x - y)},
      {x, 0, 1.5}, {y, 0, 1.5}, ScaleFunction → (# + 5 &),
      PlotPoints → 18, Frame → True, FrameLabel → {x, y}]
```

The result is shown in Figure 13.2. The option `PlotPoints → 18` increases slightly the number of vectors plotted in each direction.

This graph gives us a clearer indication of the directions of the vectors. We can find the critical points by solving the simultaneous equations $x' = 0$ and $y' = 0$ with `Solve`:

```
Solve[{x (1 - x - y) == 0, y (0.75 - 0.5 x - y) == 0}, {x, y}]
```

$$\{\{x \to 0., y \to 0.\}, \{x \to 0., y \to 0.75\}, \{x \to 0.5, y \to 0.5\},$$
$$\{x \to 1., y \to 0.\}\}$$

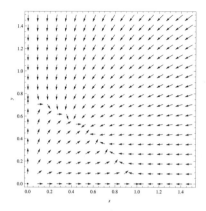

Figure 13.2: Improved Vector Field for System (13.2)

Thus the critical points are $(0,0)$, $(0,0.75)$, $(0.5,0.5)$, and $(1,0)$. Knowing the critical points, we can deduce from the vector field that $(0,0)$ is an unstable node, $(0,0.75)$ and $(1,0)$ are unstable saddle points, and $(0.5,0.5)$ is an asymptotically stable node. Solutions starting near the saddle points (but not on the axes) tend away from them, and those starting near the point $(0.5,0.5)$ tend toward it. Furthermore, the vector field strongly suggests that every solution curve starting in the first quadrant (but not on the axes) tends toward $(0.5,0.5)$. Hence $(0.5,0.5)$, which corresponds to equal populations, is apparently the limiting state that all positive solutions approach as t increases. We have thus answered both questions (a) and (b) from the beginning of this chapter on the basis of the vector field. Figure 13.3 provides a closer look at the vector field near the point $(0.5,0.5)$.

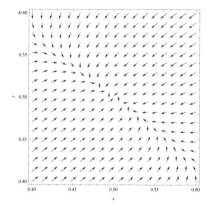

Figure 13.3: Blow-up of Vector Field for System (13.2)

Now consider the predator-prey with logistic growth system (13.3). Here $x(t)$ represents the population of prey and $y(t)$ the population of predators. The critical points of this system are $(0,0)$, $(5,0)$, and $(2,3)$. As in system (13.2), our goal is to use the vector field to answer questions (a) and (b). Here are the appropriate *Mathematica* commands:

```
VectorFieldPlot[{x (5 - x - y), y (- 2 + x)}, {x, 0, 6},
      {y, 0, 5}, ScaleFunction → (# + 75 &),PlotPoints → 18,
      Frame → True, FrameLabel → {x, y}]
```

This time we used the constant 75 in the scaling function. The output is shown in Figure 13.4.

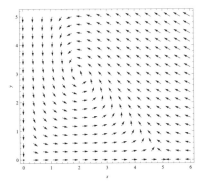

Figure 13.4: Vector Field for System (13.3)

It appears that $(0,0)$ and $(5,0)$ are unstable saddle points and that solutions spiral around $(2,3)$. The vectors near $(2,3)$ suggest that the solutions spiral into $(2,3)$, and linearized stability analysis (*cf.* the solution to Problem 6 in Problem Set F in the *Sample Solutions*) confirms that $(2,3)$ is an asymptotically stable spiral point. In particular, most solutions starting near the first two critical points tend away from them, whereas solution curves starting near the latter tend toward it. Furthermore, the vector field strongly suggests that every solution curve starting in the first quadrant (but not on the axes) tends toward the critical point $(2,3)$. This is the strongest kind of stable equilibrium—namely, no matter what the initial populations are, as long as they are both positive, the population tends toward the equilibrium position of 60% predators and 40% prey.

Remarks

- We indicated above that vector fields could be drawn by hand. It would be difficult, however, to draw as satisfactory a vector field as we have produced with *Mathematica*. On the other hand, we note that careful study of the signs of F and G in system (13.1) can lead to a good understanding of the nature of the vector field.

- In Chapter 6 we used a command of the form

  `VectorFieldPlot[{1, h[x, y]}, {x, x0, x1}, {y, y0, y1}]`

  to plot the direction field of the single first order differential equation

  $$\frac{dy}{dx} = h(x, y).$$

  This corresponds to plotting the vector field of the system

  $$\begin{cases} x' = 1 \\ y' = h(x, y). \end{cases}$$

  The "1" here explains the **1** that was used in the command for plotting direction fields.

- The essential idea in Chapter 6 for a single autonomous first order equation

  $$x' = f(x)$$

  is to determine the qualitative nature of the solutions purely on the basis of the sign of $f(x)$ for various x's. Actually, we have done the same thing here: namely, we have determined the qualitative nature of the solutions (trajectories) of a system from information on the signs of F and G.

- As mentioned above, according to linearized stability analysis, the nature of critical points for nonlinear systems can be deduced from the associated linear systems. If a critical point is asymptotically stable for the associated linear system (the case if the eigenvalues are negative or have negative real parts), then the point is asymptotically stable for the original nonlinear system. If it is unstable for the linear system, it is unstable for the nonlinear system. Centers are ambiguous. This analysis is valid in exactly the same way for a single equation. Suppose $\bar{x}$ is a critical point for $x' = f(x)$. Then letting $u(t) = x(t) - \bar{x}$, we see that

  $$u' = f(u + \bar{x}) \approx f(\bar{x}) + f'(\bar{x})u = f'(\bar{x})u,$$

  so the associated linear equation is $u' = f'(\bar{x})u$. The solutions are $u(t) = ce^{\lambda t}$, where $\lambda = f'(\bar{x})$. Therefore, $x(t) \approx \bar{x} + ce^{\lambda t}$. It is clear that $\bar{x}$ is asymptotically stable if $\lambda < 0$ and unstable if $\lambda > 0$. The same is true of the original nonlinear equation (with the situation for $\lambda = 0$ being ambiguous). Thus we see that this assessment of stability parallels the situation for systems.

Conclusions

You can calculate solutions of nonlinear systems and plot their trajectories with **NDSolve**. We described ways to do this in Chapter 12. Taken together, the information provided by a vector field, by a plot of the trajectories, and by linearized stability analysis gives a fairly complete understanding of the behavior of the solutions of a system of two first order equations. The information provided by these three approaches is similar, but each approach yields information the other approaches don't provide. For example, a vector field provides a good global indication of what the phase portrait will look like, and indicates the approximate locations of the critical points and their types; a plot of the trajectories gives a depiction of specific solution curves; and a linearized stability analysis determines the type of each critical point.

In using these approaches to study a system, it is almost always best to start with a vector field plot or with linearized stability analysis. Both approaches provide exact information about the solution curves, and a vector field does so with particularly simple calculations. By using the information from a vector field plot or from linearized stability analysis, you can choose appropriate initial data points to use for a phase portrait.

Problem Set F

Systems of Differential Equations

The solution to Problem 6 appears in the *Sample Solutions*. In Chapters 12 and 13 you can find a number of suggestions for using *Mathematica* to solve these problems. The parameters in those suggestions may need to be modified according to the problem at hand. In particular, you may have to change the time interval and/or the set of initial conditions. Also, be aware that *Mathematica* may take a long time to generate phase portraits. You should start with a relatively short time interval and a few initial conditions to get an idea of what the phase portrait looks like. Then, based on what you see, you should increase the time interval and/or the number of initial conditions as appropriate to improve the picture.

1. In this problem, we study three systems of equations taken from Boyce and DiPrima:

$$\mathbf{x}' = \begin{pmatrix} 2 & -1 \\ 3 & -2 \end{pmatrix} \mathbf{x} \qquad \text{(Prob. 3, Sect. 7.5)},$$

$$\mathbf{x}' = \begin{pmatrix} 1 & -1 \\ 5 & -3 \end{pmatrix} \mathbf{x} \qquad \text{(Prob. 5, Sect. 7.6)},$$

$$\mathbf{x}' = \begin{pmatrix} -3 & 5/2 \\ -5/2 & 2 \end{pmatrix} \mathbf{x} \qquad \text{(Prob. 4, Sect. 7.8)}.$$

(a) Use *Mathematica* to find the eigenvalues and eigenvectors of each linear system. Use your results to write down the general solution of the system.

(b) For each system, find the general solution using **DSolve**, and plot several trajectories. On your graphs, draw the eigenvectors (if relevant), and indicate the direction of increasing time on the trajectories. State the type and stability of the origin as a critical point. You may find that the quality of your portraits can be enhanced by modifying the t-interval or the range of values assumed by the initial data.

(c) For the first system only, compare the general solution you obtained in part (a) with the solution generated by **DSolve**. Are they the same? If not, explain how they are related (that is, match up the constants).

2. Do Problem 1 but replace the three systems listed there by the following three from Boyce and DiPrima:

$$\mathbf{x}' = \begin{pmatrix} -2 & 1 \\ 1 & -2 \end{pmatrix} \mathbf{x} \qquad \text{(Prob. 5, Sect. 7.5)},$$

$$\mathbf{x}' = \begin{pmatrix} 2 & -5/2 \\ 9/5 & -1 \end{pmatrix} \mathbf{x} \qquad \text{(Prob. 4, Sect. 7.6)},$$

$$\mathbf{x}' = \begin{pmatrix} 3 & -4 \\ 1 & -1 \end{pmatrix} \mathbf{x} \qquad \text{(Prob. 1, Sect. 7.8)}.$$

3. Use **Eigensystem** to find the eigenvalues and eigenvectors of the following systems. Use the eigenvalues and eigenvectors to write down the general solutions.

(a)

$$\mathbf{x}' = \begin{pmatrix} 3 & -1 \\ 1 & -2 \end{pmatrix} \mathbf{x}.$$

Let $\mathbf{x}(t) = (x(t), y(t))$. Determine the possible limiting behavior of $x(t)$, of $y(t)$, and of $x(t)/y(t)$ as t approaches $+\infty$.

(b)

$$\mathbf{x}' = \begin{pmatrix} 3 & 3 \\ -3 & -2 \end{pmatrix} \mathbf{x}.$$

(c)

$$\mathbf{x}' = \begin{pmatrix} -2 & -1 & 2 \\ 0 & 4 & 5 \\ 0 & -1 & 0 \end{pmatrix} \mathbf{x}.$$

Find the solution with initial condition

$$\mathbf{x}(0) = \begin{pmatrix} 7 \\ 5 \\ 5 \end{pmatrix}.$$

(d) Now solve the initial value problem in (c) with **DSolve** and compare the answer with the solution you obtained in (c).

4. *Mathematica* can solve some inhomogeneous linear systems. Here are three such systems (taken from Boyce and DiPrima, Sect. 7.9, Problems 2, 5, and 11):

$$\mathbf{x}' = \begin{pmatrix} 1 & \sqrt{3} \\ \sqrt{3} & -1 \end{pmatrix} \mathbf{x} + \begin{pmatrix} e^t \\ \sqrt{3}e^{-t} \end{pmatrix}, \qquad \text{(F.1)}$$

$$\mathbf{x}' = \begin{pmatrix} 4 & -2 \\ 8 & -4 \end{pmatrix} \mathbf{x} + \begin{pmatrix} t^{-3} \\ -t^{-2} \end{pmatrix}, \tag{F.2}$$

$$\mathbf{x}' = \begin{pmatrix} 2 & -5 \\ 1 & -2 \end{pmatrix} \mathbf{x} + \begin{pmatrix} 0 \\ \cos t \end{pmatrix}. \tag{F.3}$$

(a) Solve the three systems using **DSolve**.

(b) Now impose the initial data

$$\mathbf{x}(t_0) = \begin{pmatrix} 1 \\ 1 \end{pmatrix}$$

on each system, where $t_0 = 0$ in (F.1) and (F.3) and $t_0 = 1$ in (F.2). For each of (F.1)–(F.3), draw the solution curve using a time interval $-1 \le t \le 1$ in (F.1) and (F.3), and the interval $\frac{1}{2} \le t \le \frac{3}{2}$ in the other case. Based on the differential equation, indicate the direction of motion through the initial value. Then expand the time interval and, purely on the graphical evidence, venture a guess as to the behavior of the solution curve as $t \to \infty$ and as $t \to -\infty$ ($t \to 0$ in (F.2)).

5. Here we reconsider the pendulum models examined in Problems 3–6 of Problem Set D. (See also the discussion in Sections 9.2 and 9.3 of Boyce and DiPrima.)

(a) Consider first the undamped pendulum

$$\theta'' + \sin \theta = 0, \quad \theta(0) = 0, \quad \theta'(0) = b.$$

Let $x = \theta$ and $y = \theta'$; then x and y satisfy the system

$$\begin{cases} x' = y \\ y' = -\sin x, \end{cases} \qquad \begin{cases} x(0) = 0 \\ y(0) = b. \end{cases}$$

Solve this system numerically and plot, on a single graph, the resulting trajectories for the initial velocities $b = 0.5, 1, 1.5, 2, 2.5$. Use a positive time range e.g., $0 \le t \le 15$.

(b) Based on your plot in part (a), describe physically what the pendulum seems to be doing in the three cases $\theta'(0) = 1.5, 2$, and 2.5.

(c) The energy of the pendulum is defined as the sum of kinetic energy $(\theta')^2/2$ and potential energy $1 - \cos \theta$, so

$$E = \frac{1}{2}(\theta')^2 + 1 - \cos \theta = \frac{1}{2}y^2 + 1 - \cos x.$$

Show, by taking the derivative dE/dt, that E is constant when $\theta(t)$ is a solution to the equation. What basic physical principle does this represent?

(d) Use `ContourPlot` to plot the level curves for the energy. Choose the x and y ranges according to where most of the solutions from part (a) lie. Explain how your graph is related to the trajectory plot from part (a).

(e) If the pendulum reaches the upright position (*i.e.*, $\theta = \pi$), what must be true of the energy? Now explain why there is a critical value E_0 of the energy, below which the pendulum swings back and forth without reaching the upright position, and above which it swings overhead and continues to revolve in the same direction. What is the critical value E_0? What is the value of b corresponding to E_0? What do you expect the pendulum to do when b has this critical value? What did the plot in part (a) show? Explain.

(f) Now consider the damped pendulum

$$\theta'' + 0.5\theta' + \sin \theta = 0, \quad \theta(0) = 0, \quad \theta'(0) = b.$$

Numerically solve the corresponding first order system (which you must determine) and plot the resulting trajectories for the initial velocities $b = 0.5$, 1, 1.5, ... , 6. You will notice that each of the trajectories in your graph tends toward either the critical point $(0,0)$ or the critical point $(2\pi, 0)$. Explain what this means physically. There is a value b_0 for which the trajectory tends toward the critical point $(\pi, 0)$. Estimate, up to two-decimal accuracy, the value of b_0. What would this correspond to physically?

(g) Recall the energy defined in part (c). Compute E' and determine which of the following are possible: (1) E is increasing, (2) E is decreasing, (3) E is constant. Explain how the possibilities are reflected in the solutions to part (f).

6. Consider the *competing species* model (Boyce and DiPrima, Prob. 2, Sect. 9.4)

$$\begin{cases} \frac{dx}{dt} = x(1.5 - x - 0.5y) \\ \frac{dy}{dt} = y(2 - 1.5x - 0.5y) \end{cases}$$

for $x, y \geq 0$.

(a) Find all critical points of the system. At each critical point, calculate the corresponding linear system and find the eigenvalues of the coefficient matrix; then identify the type and stability of the critical point.

(b) Plot the vector field on a region small enough to distinguish the critical points but large enough to judge the possible solution behaviors away from the critical points.

(c) Use several initial data points (x_0, y_0) in the first quadrant to draw a phase portrait for the system. Identify the direction of increasing t on the trajectories you obtain. Use the information from parts (a) and (b) to choose a representative sample of initial conditions. Then combine the vector field and phase portrait on a single graph.

(d) Suppose the initial state of the population is given by

$$x(0) = 2.5, \ y(0) = 2.$$

Find the state of the population at $t = 1, 2, 3, 4, 5, \ldots, 20$.

(e) Explain why, practically speaking, there is no "peaceful coexistence"; *i.e.*, with the exception of an atypical set of starting populations (the *separatrix* curves), one or the other population must die out. For which nonzero initial populations is there no change? Add to your final plot from part (c) the separatrices, *i.e.*, the solution curves that approach the unstable equilibrium point where both populations are positive. These curves form the boundary between the solution curves that approach each of the two stable points where only one population survives. *Hint*: Since a separatrix is a solution curve that *approaches* a saddle point, you can obtain it by plotting trajectories starting very close to the saddle point and going *backwards* in time.

(f) The vertical line $x = 1.5$ cuts the separatrix. By using the hint in part (e) and the event detection feature of **NDSolve** (see Chapter 8), find a numerical approximation to the number $\bar{y}$ such that $(1.5, \bar{y})$ is on the separatrix. Could you verify numerically whether the point is in fact on the separatrix? Why or why not?

7. Consider the *competing species* model (Boyce and DiPrima, Prob. 4, Sect. 9.4)

$$\begin{cases} \frac{dx}{dt} = x(1.5 - 0.5x - y) \\ \frac{dy}{dt} = y(0.75 - 0.125x - y). \end{cases}$$

for $x, y \geq 0$.

(a) Find all critical points of the system. At each critical point, calculate the corresponding linear system and find the eigenvalues of the coefficient matrix; then identify the type and stability of the critical point.

(b) Plot the vector field on a region small enough to distinguish the critical points but large enough to judge the possible solution behaviors away from the critical points.

(c) Use several initial data points (x_0, y_0) in the first quadrant to draw a phase portrait for the system. Identify the direction of increasing t on the trajectories you obtain. Use the information from parts (a) and (b) to choose a representative sample of initial conditions. Then combine the vector field and phase portrait on a single graph.

(d) Suppose the initial state of the population is given by

$$x(0) = 0.1, \ y(0) = 0.1.$$

Find the state of the population at $t = 1, 2, 3, 4, 5, \ldots, 20$.

(e) Explain why, practically speaking, "peaceful coexistence" is the only outcome; *i.e.*, with the exception of the situation in which one or both species starts out without any population, the population distributions always tend toward a certain equilibrium point. Sketch on your final plot from part (c) the separatrices that connect the stable equilibrium point to the two unstable points at which one population is zero; these separatrices divide the solution curves that tend toward the origin as $t \to -\infty$ from those that are unbounded as $t \to -\infty$. *Hint*: Since the separatrices are trajectories coming *out* of the saddle points, one can plot them by solving the equation for initial conditions very close to the saddle points.

(f) The vertical line $x = 2.5$ cuts a separatrix. By using the hint in part (e) and the event detection feature of **NDSolve** (see Chapter 8), find a numerical approximation to the number $\bar{y}$ such that $(2.5, \bar{y})$ is on the separatrix.

8. Consider the *predator-prey* model

$$
\begin{cases}
\frac{dx}{dt} = x(4 - 3y) \\
\frac{dy}{dt} = y(x - 2)
\end{cases}
$$

in which $x \geq 0$ represents the population of the prey and $y \geq 0$ represents the population of the predators.

(a) Find all critical points of the system. At each critical point, calculate the corresponding linear system and find the eigenvalues of the coefficient matrix; then identify the type and stability of the critical point.

(b) Plot the vector field on a region small enough to distinguish the critical points but large enough to judge the possible solution behaviors away from the critical points.

(c) Use several initial data points (x_0, y_0) in the first quadrant to draw a phase portrait for the system. Identify the direction of increasing t on the trajectories you obtain. Use the information from parts (a) and (b) to choose a representative sample of initial conditions. Then combine the vector field and phase portrait on a single graph.

(d) Explain from your phase portrait how the populations vary over time for initial data close to the unique critical point inside the first quadrant. What happens for initial data far from this critical point?

(e) Suppose the initial state of the population is given by

$$
x(0) = 1, \ y(0) = 1.
$$

Find the state of the population at $t = 1, 2, 3, 4, 5$.

(f) Estimate to two decimal places the period of the solution curve that starts at $(1, 1)$.

9. Consider a *predator-prey* model where the behavior of the prey is governed by a logistic equation (in the absence of the predator). Such a model is typified by the following system (Boyce and DiPrima, Prob. 3, Sect. 9.5):

$$\begin{cases} \frac{dx}{dt} = x(1 - 0.5x - 0.5y) \\ \frac{dy}{dt} = y(-0.25 + 0.5x), \end{cases}$$

where $x \geq 0$ represents the population of the prey and $y \geq 0$ represents the population of the predators.

(a) Find all critical points of the system. At each critical point, calculate the corresponding linear system and find the eigenvalues of the coefficient matrix; then identify the type and stability of the critical point.

(b) Plot the vector field on a region small enough to distinguish the critical points but large enough to judge the possible solution behaviors away from the critical points.

(c) Use several initial data points (x_0, y_0) in the first quadrant to draw a phase portrait for the system. Identify the direction of increasing t on the trajectories you obtain. Use the information from parts (a) and (b) to choose a representative sample of initial conditions. Then combine the vector field and phase portrait on a single graph.

(d) Explain from your phase portrait how the populations vary over time for initial data close to the unique critical point inside the first quadrant. What happens for initial data far from this critical point?

(e) Suppose the initial state of the population is given by

$$x(0) = 1, \; y(0) = 1.$$

Find the state of the population at $t = 1, 2, 3, 4, 5$.

(f) Estimate how long it takes for both populations to be within 0.01 of their equilibrium values if we start with initial data $(1, 1)$.

10. Consider a modified predator-prey system where the behavior of the prey is governed by a logistic/threshold equation (in the absence of the predator). Such a model is typified by the following system (Boyce and DiPrima, Prob. 5, Sect. 9.5):

$$\begin{cases} \frac{dx}{dt} = x(-1 + 2.5x - 0.3y - x^2) \\ \frac{dy}{dt} = y(-1.5 + x) \end{cases}$$

where $x \geq 0$ represents the population of the prey, and $y \geq 0$ represents the population of the predators.

(a) Find all critical points of the system. At each critical point, calculate the corresponding linear system and find the eigenvalues and eigenvectors of the coefficient matrix; then identify the critical points as to type and stability.

(b) Plot the vector field on a region small enough to distinguish the critical points but large enough to judge the possible solution behaviors away from the critical points.

(c) Use several initial data points (x_0, y_0) in the first quadrant to draw a phase portrait for the system. Identify the direction of increasing t on the trajectories you obtain. Use the information from parts (a) and (b) to choose a representative sample of initial conditions. Then combine the vector field and phase portrait on a single graph.

(d) In parts (a), (b), and (c), you obtained information about the solutions of the system using three different approaches. Combine all of this information by combining the plots from parts (b) and (c), and then marking the critical points and adding any separatrices. (See the hints in Problems 6 and 7 about how to plot separatrices.) What information does the approach in part (a) provide that the other approaches do not? Answer the same question for parts (b) and (c).

(e) Interpret your conclusions in terms of the populations of the two species.

11. Some of the nonlinear systems of differential equations we have studied have more than one asymptotically stable equilibrium point. In such cases, it can be difficult to predict which (if any) equilibrium point the solution with a given initial condition will approach as time increases. Generally, the solution curves that approach one stable equilibrium will be separated from the solutions that approach another stable equilibrium by a separatrix curve. The separatrix is itself a solution curve that does not approach either stable equilibrium—often it approaches a saddle point instead. (See, for example, Figure 9.4.4 in Section 9.4 of Boyce and DiPrima.)

Having located a relevant saddle point, one can approximate the separatrix by choosing an initial condition very close to the saddle point and solving the differential equation *backwards* in time. The reason for going "back in time" is to find (approximately) a solution curve that approaches very close to the saddle point as time increases. We will apply this idea in a population dynamics model to get a fairly precise picture of which initial conditions correspond to which equilibria.

(a) Consider the system

$$\begin{cases} \frac{dx}{dt} = x(-1 + 2.5x - 0.3y - x^2) \\ \frac{dy}{dt} = y(-1.5 + x) \end{cases}$$

studied in Problem 10. There are two asymptotically stable equilibria, $(0, 0)$ and $(3/2, 5/3)$, and two saddle points at $(0.5, 0)$ and $(2, 0)$. Plot a family of solution curves in the first quadrant (positive x and y) on the same graph. Use a time interval from $t = 0$ to a positive time large enough that you can clearly see where the solutions curves are headed, but not so large that your plot takes forever to compute. Choose a set of initial values robust enough to give a reasonably "thick" set of solution curves. It also may be useful to adjust the range of the plot until you get an approximately square graph. Observe where the solution curves seem to be heading and where the separatrix appears to be.

(b) Draw an approximate separatrix by plotting a solution curve with initial values for x and y very close to the saddle point $(1/2, 0)$, using a *negative* range of values for t. This should give you a good picture of the separatrix. Superimpose the separatrix on the portrait of solution curves you drew in part (a). Make sure the separatrix really does separate the different asymptotic properties of the solution curves found in part (a). Discuss how the possible limiting values of x and y of a solution curve depend on its initial values.

12. In Section 5.3 we investigated the sensitivity to initial values of solutions of a single differential equation. In this problem, we investigate the same issue for a system of equations. The system below is said to be *chaotic* because the solutions are sensitive to initial values, but in contrast to Example 5.3 and Problems 14 and 15 of Problem Set C, the solutions remain bounded.

Since we do not know the exact solutions for the system we study, in order to judge the time it takes for a small perturbation of the system to become large, we will compare pairs of numerical solutions whose initial conditions are close to each other. We consider perturbations of decreasing size from 10^{-1} to 10^{-4}.

(a) Consider the Lorenz system

$$\begin{cases} x' = 10(y - x) \\ y' = 28x - y - xz \\ z' = -(8/3)z + xy \end{cases}$$

(which is studied, for example, in Section 9.8 of Boyce and DiPrima). We investigate the idea that small changes in the initial conditions can lead to large changes in the solution after a relatively short period of time. For $\delta = 0.1$, plot on the same graph the first coordinate $x(t)$ of the solutions corresponding to the initial conditions $(3, 4, 5)$ and $(3, 4, 5 + \delta)$ from $t = 0$ to $t = 20$. Observe the time at which the solutions start to differ substantially. Repeat for $\delta = 0.01, 0.001, 0.0001$. (The criterion for a "substantial" difference between the two solutions is up to you; just try to be consistent from one observation to the next.)

(b) Make a table or graph of the number of decimal places in which the initial condition was perturbed (that is, 1, 2, 3, and 4 for the four given values of δ) versus the amount of time the solutions stayed close to each other. This graph shows roughly how long we should trust a numerical solution to be close to the actual solution for a given number of digits of accuracy in each step of our numerical method. Describe how the amount of time we can trust a numerical solution seems to depend on the number of digits of accuracy per time step, judging from your data. If a numerical method has 14 digits of accuracy, about how long do you think the numerical solution can be trusted? Roughly how many digits of accuracy would be needed to trust the solution up to $t = 1000$?

13. Read the introduction to Problem 12.

   (a) Do part (a) of Problem 12, but using the Rössler system

$$\begin{cases} x' = -y - z \\ y' = x + 0.36y \\ z' = 0.4x - 4.5z + xz. \end{cases}$$

   This time plot $y(t)$ from $t = 0$ to $t = 100$ and use $(2, 2, 2)$ as the initial condition in place of $(3, 4, 5)$.

   (b) Do part (b) of Problem 12 for the Rössler system, this time initially assuming a numerical method accurate to 10 digits.

14. A simple model of a nonlinear network, of a sort that comes up for example in modeling portions of the nervous system, consists of a number n of identical units connected together, with the state x_j of the jth unit satisfying the differential equation

$$x'_j = f(x_j - x_{j-1}), \tag{F.4}$$

meaning that the rate of change of each unit is a fixed function f of the difference between the state of the unit and the state of its predecessor. (See Figure F.1 for a schematic diagram.) For the sake of this problem, take $x_0 = x_n$. (That means the

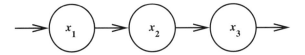

Figure F.1: A Multi-Unit Network

network closes up, with the last unit providing input to the first, so that (F.4) becomes $x'_1 = f(x_1 - x_n)$ for $j = 1$.) Suppose $n = 5$.

   (a) The simplest interesting model of this sort occurs when f is linear. To encourage equilibrium, we want x_j to increase when x_{j-1} is bigger than it, and to decrease when x_{j-1} is smaller. This suggests taking $f(x_j - x_{j-1}) = -\lambda(x_j - x_{j-1})$ with $\lambda > 0$. In other words,

$$\mathbf{x}' = \lambda A \mathbf{x}, \qquad A = \begin{pmatrix} -1 & 0 & 0 & 0 & 1 \\ 1 & -1 & 0 & 0 & 0 \\ 0 & 1 & -1 & 0 & 0 \\ 0 & 0 & 1 & -1 & 0 \\ 0 & 0 & 0 & 1 & -1 \end{pmatrix}. \tag{F.5}$$

Have *Mathematica* verify that $(A + I)^5 = I$. That says that the eigenvalues of A are of the form $-1 + e^{2\pi ik/5}$, or that the eigenvalues of λA are of the form $\lambda(-1 + e^{2\pi ik/5})$, for integers k from 0 to 4. What does this say about the solutions to (F.5) as $t \to \infty$?

(b) Set $\lambda = 1$ and solve the system (F.5) with various initial conditions to give examples that demonstrate your conclusion from part (a).

(c) Now let $f(x) = -bx^3 + x^2 + x$, choose a value of $b > 0$, and solve the system (F.4) numerically using a randomly generated initial condition. Experiment with different positive values of b, looking for values that make a qualitative difference in the behavior of solutions to (F.4) as $t \to \infty$. Discuss how the results compare with the linear model in parts (a) and (b).

15. This problem is a follow-up to Problem 14 above. One possible modification to the model in that problem is to assume that it takes a certain amount of time T for the difference between the states of a unit and its predecessor to have an effect. (For example, if the model represents a network of neurons, it might take time for signals to propagate across the synapse between them.) For simplicity, let's consider the linear model in (F.5). The matrix A remains unchanged, but the equation now becomes a differential/difference equation '

$$\mathbf{x}'(t) = \lambda A\mathbf{x}(t - T). \tag{F.6}$$

We can still find solutions to this equation of the form $\mathbf{x}(t) = e^{\sigma t}\mathbf{v}$ provided that $\mathbf{v}$ is an eigenvector of A. For if $A\mathbf{v} = \mu\mathbf{v}$, then substituting in (F.6) gives a *scalar* equation relating λ, σ, μ, and T.

(a) Find this equation. (Note that σ, μ, and $\mathbf{v}$ are allowed to be complex; this corresponds to allowing solutions involving sine and cosine functions after we take real parts.)

(b) A *standing wave* solution to (F.6) is a purely oscillatory solution, *i.e.*, one for which σ is purely imaginary. Show that such solutions exist for certain special values of the parameters.

(c) Let $T = 1$ and demonstrate what happens to a typical solution of (F.6) as λ increases from 0 to 1.

16. The subject of *chemical kinetics* studies the rate at which chemical reactions take place. For a reaction that takes place in one step, say

$$A + B \to AB,$$

the rate at which the reaction proceeds to the right (at least at low concentrations of the reactants) is proportional to the product of the concentrations of A and B, since formation of compound AB depends on the collision of a molecule of A with a molecule of B. (Strictly speaking, we are talking about a situation where all the substances involved are gases, and it makes sense to talk about molecular collisions. The case where A and B are dissolved in water is slightly more complicated.)

But for a reaction like
$$2A + B \to A_2 B,$$
there are two possibilities. It might be that the reaction proceeds in one step, in which case the rate would be proportional to the product of the *square* of the concentration of A with the concentration of B. Such a reaction is called *second order* in A, *first order* in B. Or it might be that the reaction really proceeds in two steps,
$$A + B \to AB, \qquad A + AB \to A_2 B,$$
in which case the square of the concentration of A never comes in (since there is no need for two molecules of A to collide). In this problem, we analyze the second possibility. Let x, y, z, and w be the concentrations of A, B, AB, and $A_2 B$, respectively (in units proportional to the total number of molecules). Conservation of matter says that
$$x + z + 2w \quad \text{(total amount of } A\text{)}, \qquad y + z + w \quad \text{(total amount of } B\text{)}$$
remain constant. For definitiveness, we will assume that (again, in appropriate units)
$$x + z + 2w = 2, \qquad y + z + w = 1. \tag{F.7}$$

(a) Use **Solve** to solve the equations (F.7) for w and z. Your result should be $w = 1 - x + y$, $z = x - 2y$.

(b) The two-step chemical reaction is modeled by the differential equations
$$w' = \beta x z$$
(rate of creation of $A_2 B$ is proportional to xz)
$$z' = \alpha x y - \beta x z$$
(AB is being created from A and B, destroyed by creation of $A_2 B$)

for positive constants α and β. Use part (a) and **Solve** to rewrite the differential equations in terms of x, y, x' and y'. This should yield
$$\begin{cases} x' = (2\beta - \alpha)xy - \beta x^2 \\ y' = -\alpha xy, \end{cases} \tag{F.8}$$
an autonomous system that falls into the framework of Chapter 13.

(c) Show that there is only one critical point of (F.8) in the physically realistic domain where x, y, z, and w are all ≥ 0. What is it? Explain why that critical point does not fall under the rubric of *almost linear system* theory, as developed for example in Boyce and DiPrima, Section 9.3.

(d) Nevertheless, we can uncover a great deal of qualitative information about the solution curves. Plot the vector field in the case $\alpha = 1$, $\beta = 10$. (This is the case where the intermediate product AB is relatively unstable and converts rapidly to A_2B.) What does the field look like in the physically relevant region? Superimpose several trajectories of the system on your plot, starting along the line segment joining $x = 1$, $y = 0$ and $x = 2$, $y = 1$. How do the solution curves approach the critical point?

(e) Plot the vector field in the case $\alpha = 10$, $\beta = 1$. (This is the case where the intermediate product AB is relatively stable.) Again superimpose several trajectories of the system on your plot, starting along the line segment joining $x = 1$, $y = 0$ and $x = 2$, $y = 1$. How do the results differ from the previous case?

(f) Finally, suppose the initial state of the system is $x = 2$, $y = 1$ (corresponding to $w = z = 0$, *i.e.*, no AB or A_2B). Plot the concentration w of the final product A_2B as a function of time t, in the two cases studied above ($\alpha = 1$ and $\beta = 10$, then $\alpha = 10$ and $\beta = 1$), on the same set of axes. Comment on the difference. How do the results compare with what you would get in the case where the reaction

$$2A + B \rightarrow A_2B$$

proceeds in a single step?

17. This problem continues the discussion of chemical kinetics in Problem 16. Again consider a gaseous reaction, but this time assume that it is reversible, so that we have a reaction of the form

$$A + B \rightleftharpoons AB.$$

Let x, y, and z be the concentrations of A, B, and AB, respectively. Just as in Problem 16, $x + z$ and $y + z$ remain constant in time, because of conservation of matter. Let's assume that (in suitable units) $x + z = 1$, but that $y + z$ can be bigger than 1. Since AB is being both created (at a rate proportional to the product of the concentrations of A and B) and destroyed (at a rate proportional to the concentration of AB), the differential equation becomes:

$$z' = -x' = -y' = \alpha xy - \gamma z = \alpha xy - \gamma(1 - x),$$

or

$$\begin{cases} x' = -\alpha xy + \gamma(1 - x), \\ y' = -\alpha xy + \gamma(1 - x). \end{cases} \tag{F.9}$$

Study this system in the domain $0 \le x \le 1, y \ge 0$.

(a) Show that (F.9) has a curve of critical points given by the equation

$$y = \frac{\gamma}{\alpha}\left(\frac{1}{x} - 1\right).$$

(b) Plot the vector field in the case $\alpha = 1$, $\gamma = 10$. (This is the case where the product AB is relatively unstable and converts back rapidly to A and B.) What does the field look like in the physically relevant region? Superimpose several trajectories of the system on your plot, starting along the line segment joining $x = 0$, $y = 0$ and $x = 0$, $y = 3$. (*Note:* $x = 0$ means no A is present initially, but then $z = 1$, so AB is present, and because of the reversible reaction, A is eventually created.)

(c) Plot the vector field in the case $\alpha = 10$, $\gamma = 1$. (This is the case where the product AB is relatively stable.) Again superimpose several trajectories of the system on your plot, starting along the line segment joining $x = 0$, $y = 0$ and $x = 0$, $y = 3$. How do the results differ from the previous case?

(d) Finally, suppose the initial state of the system is $x = 0$, $y = 2$. Plot the concentration z of the product AB as a function of time t, in the two cases studied above ($\alpha = 1$ and $\gamma = 10$, then $\alpha = 10$ and $\gamma = 1$), on the same set of axes. What do you observe?

18. A magnetic servomotor is depicted in Figure F.2. The input voltage $v(t)$ results in a current $i(t)$, which produces a torque on the rotor. The rotor, with angle $\theta(t)$, acts like a generator, and a circuit-torque analysis leads to the pair of differential equations (*cf.* E. B. Magrab, S. Azarm, B. Balachandran, J. Duncan, K. Herold, G. Walsh, *An Engineer's Guide to MATLAB*, Prentice Hall, 2001)

$$\begin{cases} Li'(t) + k_b\theta'(t) + Ri(t) = v(t) \\ J\theta''(t) + b\theta'(t) - k_\tau i(t) = 0, \end{cases} \tag{F.10}$$

where the constants are: R (motor resistance), L (inductance), J (inertia), b (motor friction), k_b (back emf generator constant), and k_τ (conversion factor from current to torque).

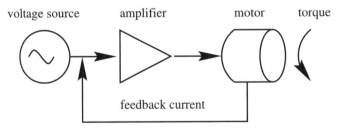

voltage source   amplifier   motor   torque

feedback current

Figure F.2: Schematic Picture of a Magnetic Servomotor

(a) Convert the pair of equations to a system of three first order linear equations.

(b) Assign values to the constants as follows (we ignore units here): $R = 5.0$, $L = 0.005$, $J = 0.03$, $b = 0.01$, $k_b = 0.125$, $k_\tau = 15$. Now solve the system using

DSolve. For initial data you may take $i(0) = \theta(0) = \theta'(0) = 0$ and set the impressed voltage to be $v(t) \equiv 1$. What is the motor doing? (To answer, graph the rotor angle $\theta(t)$.) Is that realistic?

(c) Now assume there is a feedback mechanism. This means that

$$v(t) = \kappa(r(t) - \theta(t)),$$

where κ and $r(t)$ are to be specified. Take $r(t) \equiv 1$, the same vanishing initial data and $\kappa = 1$; then solve the resulting system. Now what is the motor doing? Is this realistic?

(d) Increase the value of κ to 50, then 150, and finally 500. Describe the different behaviors that you see and what they imply for the motor.

(e) A somewhat more realistic, albeit nonlinear, model is obtained if the feedback function is

$$v(t) = \tanh(\kappa(r(t) - \theta(t))).$$

Repeat part (d) with $\kappa = 1$ and 150. This time, instead of **DSolve**, you will need to use **NDSolve**. What happens in the second case is quite subtle. Can you identify the long-term behavior?

Glossary

This glossary is divided into the following sections:

1. *Mathematica* Operators: the special symbols used by *Mathematica*,

2. *Mathematica* Commands: commands that manipulate data or expressions, or that initiate a process,

3. Built-in Functions: basic mathematical functions that can be evaluated or plotted,

4. Graphics Commands and Options: commands and options using in creating or modifying figures,

5. Built-in Constants,

6. *Mathematica* Programming: commands used for programming,

7. *Mathematica* Packages: the *Mathematica* packages relevant to the problems in this book.

The distinction among these various categories, especially among commands and programming constructs, is somewhat artificial.

Each item is followed by a brief description of its effect and then (in most cases) one or more examples. To get more information on a command, you can use the **?** operator or the Documentation Center. This glossary is not a comprehensive list of *Mathematica* commands, but it includes the commands most useful for studying differential equations. Please keep in mind that *Mathematica* is case-sensitive and that all built-in *Mathematica* commands begin with a capital letter.

Mathematica Operators

/   Ordinary division.

/.   Replacement operator.

   a /. a → b

//   "Postfix" operator for displaying output in a given form.

   Table[N[Log[n]], {n, 1, 5}] // TableForm

`->`	Used in defining transformation rules or entering options to commands. You can also use $\rightarrow$ from the palette.

`{x, x, x} /. x → RandomReal[]`

`:>`	Defines a *delayed* transformation rule. You can also use $:\rightarrow$ from the palette.

`{x, x, x} /. x :→ RandomReal[]`

`*`	Multiplication.
`^`	Used for exponentiation.
`_`	Underscore. Used to designate arguments in a function definition.

`f[x_, y_] := x^2 + y^2`
`f[3, 4]`

`"`	Used to begin and end strings.

`"This is a string."`

`<<`	File input operator, used for loading packages.
`` ` ``	Used for the context associated to a package. Thus this character usually appears at the end of a command for loading a package.

`` <<VectorFieldPlots` ``

`'`	Differentiation operator.

`Cos'[x]`

`;`	Suppresses output of a *Mathematica* command.

`Table[i, {i, 1, 100}];`

`:=`	Delayed assignment operator.
`=`	Immediate assignment operator.
`==`	Used to define an *equation*.
`?`	Gives short version of information on a command.

`?Solve`

`%`	The last result generated. If followed by a number **n**, the output of the line labeled Out[n].
`#`	Placeholder for the argument of a pure function.
`&`	Used to define a pure function.

`f = (#^2 + 1 &); f[2]`

`||`	Logical "or" operator.

`1<2 || 1>2`

`&&`	Logical "and" operator.

`1<2 && 1>2`

Mathematica Commands

Clear   Clears values and definitions for variables and functions. If you specify one or more variables, then only those variables are cleared.

    f = 10; Clear[f, g]

Collect   Collects coefficients of powers of the specified symbolic variable in a given symbolic expression.

    Collect[x^2 - 2*x^2 + 3*x + x*y, x]

ComplexExpand Expands expressions with complex exponentials.

    ComplexExpand[Exp[2 + 5I]]

Conjugate Gives the complex conjugate of a complex number.

    Conjugate[2 + 3I]

ConstantArray   Creates a matrix with all entries the same.

    ConstantArray[c, {3, 4}]

D   Differentiation or partial differentiation operator. You can also use the ∂ operator from the palette.

    D[x*y, x]

Det   The determinant of a square matrix.

    Det[{{1, 3}, {4, 5}}]

DSolve   Symbolic ODE solver.

    DSolve[y''[x] + x*y[x] == 0, y[x], x]

Eigenvalues   Computes eigenvalues of a square matrix.

    Eigenvalues[{{x^2, 1}, {x - 1, 1}}]

Eigensystem   Computes eigenvalues and eigenvectors of a square matrix.

    Eigensystem[{{2, 1}, {1, 1}}]

Evaluate   Evaluates an expression. Useful in plotting commands.

    Plot[Evaluate[Table[x^n, {n, 0, 10}]], {x, 0, 1}]

Expand   Expands products and powers in an algebraic expression.

    Expand[(x + y)^3]

Factor   Factors a polynomial.

    Factor[x^4 - y^4]

FactorInteger Factors an integer.

    FactorInteger[99999999]

FindRoot   Tries to find a numerical solution to an equation near the given starting point. If right-hand side of equation is missing, it is assumed to be 0.

    FindRoot[Cos[x] + x, {x, 0}]

First   Selects the first element in a list.

    First[{2, 4, 6}]

```
    First[Solve[x^2 - 4 == 0, x]]
```

Flatten   Removes extra levels of braces in a nested list.
```
    Flatten[{{1, 2}, 3, {4, 5, 6, {7}}}]
    Flatten[{{1, 2}, 3, {4, 5, 6, {7}}}, 1]
```

IdentityMatrix   The identity matrix of the specified size.
```
    IdentityMatrix[5] // MatrixForm
```

Integrate   Integration operator for both definite and indefinite integrals. You can also use ∫ from the palette.
```
    Integrate[1/(1 + x^2), x]
    Integrate[Exp[-x^2], {x, 0, Infinity}]
```

Inverse   Inverse of a square matrix.
```
    Inverse[{{1 2}, {3 5}}]
```

InverseLaplaceTransform   Computes the Inverse Laplace Transform.
```
    InverseLaplaceTransform[1/s^2, s, t]
```

Join   Combines two lists.
```
    Join[{1, 2, 3}, {4, 5}]
```

LaplaceTransform Computes the Laplace Transform.
```
    LaplaceTransform[t^5, t, s]
```

Last   Selects the last element in a list.
```
    Last[{3, 4, 5}]
```

Length   Returns the number of elements in a list, or the number of rows in a matrix.
```
    Length[{{a, b}}]
```

Limit   Finds a two-sided limit, if it exists. Set the **Direction** to ± 1 for one-sided limits.
```
    Limit[Sin[x]/x, x → 0]
    Limit[Exp[1/x], x → 0, Direction → 1]
```

Map   Applies a function to each element of a list. Can also be entered as /@.
```
    Map[Sqrt, {1, 2, 3}]
```

MatrixForm Prints a nested list as a matrix.
```
    MatrixForm[{{a, b}, {c, d}}]
```

Max   Gives the largest number in a list.
```
    Max[Table[Sin[x], {x, 100}]]
```

Method   Option for **NDSolve**, asking it to use a specific method. Some of the possibilities are **"ExplicitEuler"**, **"ExplicitRungeKutta"**, and **"EventLocator"**. The latter is used to locate the point where some function of the variables becomes zero.
```
    NDSolve[{y''[t] + Sin[y[t]] == 0, y[0] == 0, y'[0] == 1}, y[t],
    {t, 0, 2}, Method → {"EventLocator", "Event" → y'[t]}]
```

`Min`   Gives the smallest number in a list.

```
Min[Table[Sin[x], {x, 100}]]
```

`N`   Numerical approximation operator.

```
N[Pi, 30]
```

`NDSolve`   Numerical ODE solver. Sometimes you may want to adjust options such as `AccuracyGoal`, `WorkingPrecision`, and `MaxSteps`. Also works with higher-order equations and systems.

```
sol = NDSolve[{y'[t] ==y[t]^2 + t, y[0] ==1}, y, {t,0,2}]
    y[0.9] /. First[sol]
sol1 = NDSolve[{x'[t] ==x[t]*(-1+2x[t] -y[t] - x[t]^2),
    y'[t] == y[t]*(-1 + x[t]), x[0] == 2, y[0] == 2},
    {x[t], y[t]}, {t, -5, 15},
    WorkingPrecision → 20, AccuracyGoal → 8];
ParametricPlot[Evaluate[{x[t],y[t]} /. First[sol1]],
    {t, -5, 15}, PlotRange → {{0, 20}, {0, 20}}]
```

`NestList`   Applies a function iteratively.

```
NestList[f, x, 3]
```

`NIntegrate` Numerical integration operator. Similar to `Integrate` followed by `N`, but may give slightly different results.

```
NIntegrate[Sqrt[1 - x^6], {x, 0, 1}]
```

`Off`   Suspends printing of an error or a warning message.

```
Off[NDSolve::ndsz]
```

`On`   Turns on printing of an error or a warning message.

```
On[NDSolve::ndsz]
```

`Print`   Displays a string or the value of a variable.

```
Print["The answer is:  ", a]
```

`Quit`   Terminates the *Mathematica* kernel.

`Remove`   Removes definitions; useful for removing definitions of names accidentally invoked before a package is loaded.

```
Remove[VectorFieldPlot]
```

`Root`   Represents a root of a polynomial equation.

```
N[Root[#1^3 - 3#1 + 1 &, 3]]
```

`Series`   Gives the series expansion of an expression around a point out to a specified order.

```
Series[Tan[x], {x, 0, 10}]
```

`SetDirectory`   Makes the specified directory the current (working) directory.

```
SetDirectory["C:\mydocs"]
```

`Simplify`   Simplifies an algebraic expression.

```
Simplify[(x+1)^2 - x^2]
```

Solve   Solves an algebraic equation. Will guess what the independent variable is
              if it is not specified.
      `Solve[x^2 + 3*x + 1 == 0]`

Sum   Computes sums, possibly infinite. You can also use $\sum$ from the palette.
      `Sum[1/n^4, {n, 1, Infinity}]`

Table   Generates a list.
      `Table[1/k, {k, 1, 5}]`

TableForm Prints a list in the form of a table.
      `TableForm[{{a, b}, {c, d}}]`

Timing   Runs a command and returns the time it took in seconds as well as the
              output.
      `Timing[Integrate[x*(x^2 + 1)^100, x]]`

Transpose Transpose of a matrix.
      `Transpose[{{1, 2}, {3, 4}}]`

Union   Amalgamates lists, sorting and eliminating duplications.
      `Union[{a, d}, {b, c, d}]`

Built-in Functions

Abs $|x|$.

AiryAi   The solution $\mathrm{Ai}(x)$ to Airy's equation $y'' - xy = 0$.

AiryBi   The solution $\mathrm{Bi}(x)$ to Airy's equation.

ArcSin $\arcsin x$.

ArcTan $\arctan x$. When used with two arguments x and y, gives the polar coordi-
                nate θ of the point (x, y) in the plane.
      `ArcTan[-1, -1]`

BesselJ   Bessel function of the first kind. `BesselJ[n, x]` is a solution, $J_n(x)$,
                to Bessel's equation of order n, $y'' + \frac{1}{x}y' + \left(1 - \frac{n^2}{x^2}\right)y = 0$.

BesselY   Bessel function of the second kind. `BesselY[n, x]` is another solution,
                $Y_n(x)$, to Bessel's equation of order n, with a logarithmic singularity at 0.

Cos $\cos x$.

Cosh $\cosh x$.

DiracDelta The Dirac δ "function" of mass 1.

Erf   The *error function* $\mathrm{erf}(x) = (2/\sqrt{\pi}) \int_0^x e^{-t^2}\, dt$.

Erfi $-i\, \mathrm{erf}(ix)$.

Exp e^x.

Gamma $\Gamma(x) = \int_0^\infty t^{x-1}e^{-t}\,dt.$

HeavisideTheta The Heaviside function with a jump at 0. Almost the same as **UnitStep** except that not defined at 0.

Im $\mathrm{Im}(z)$, the imaginary part of a complex number.

Log The natural logarithm $\ln x = \log_e x.$

Re $\mathrm{Re}(z)$, the real part of a complex number.

Sin $\sin x.$

Sinh $\sinh x.$

SinIntegral The *sine integral* $\mathrm{Si}(x) = \int_0^x (\sin t/t)\,dt.$

Sqrt $\sqrt{x}.$

Tan $\tan x.$

Tanh $\tanh x.$

UnitStep The unit step function $u(t) = 1$ for $t \geq 0$, 0 for $t < 0.$

Graphics Commands and Options

AspectRatio The ratio of height to width for a plot.
```
Table[Graphics[Circle[],AspectRatio→k], {k,.5,1.5,.5}]
```

Axes Controls whether or not to display axes.

AxesLabel Controls the labels for the coordinate axes.

AxesOrigin The point at which the axes should cross. The default, **Automatic**, does not always put this at the origin.

ContourPlot Plots level curves of a function of two variables. The option **Contours** specifies either the number of contours or the specific contours to display. The option **ContourShading** determines whether or not the regions between contours should be shaded. The option **ContourStyle** determines the appearance of the contour lines.
```
ContourPlot[Cos[x]*Sin[y], {x, -Pi, Pi}, {y, -Pi, Pi}]
ContourPlot[y^2 == x*(x^2-1), {x, -3, 3}, {y, -3, 3}]
ContourPlot[y^2 - x^3, {x, -3, 3}, {y, -3, 3},
          Contours → {-1, 0, 1}, ContourShading →
          False, ContourStyle → Thick]
```

Export Exports a plot to a file in appropriate format.
```
Export["sinplot.gif", Plot[Sin[x], {x, -Pi, Pi}]]
```

Graphics The most basic graphics command. Used to place various geometric objects in the x-y plane.
```
Graphics[{Circle[{0,0},2], PointSize[.02], Point[{0,0}],
```

```
                    Line[{{-1,0},{1,0}}]}]
```

GraphicsGrid Arranges several images into a grid of smaller plots.
```
     pp[n_, m_] := Plot[x^n, {x, 0, 1},
         PlotStyle → RGBColor[m, 0, 1 - m]]
     GraphicsGrid[Table[Table[pp[n, m], {n,1,3}],
         {m,0,1,.25}]]
```

ListPlot   Plots individual points rather than curves.
```
     ListPlot[Table[{x, Sin[x]}, {x, -Pi, Pi, Pi/10}]]
```

LogPlot   Produces a logarithmic plot.
```
     LogPlot[Cosh[x], {x, -10, 10}]
```

ParametricPlot Produces a parametric plot. Useful for plots of trajectories of ODE
                systems.
```
     ParametricPlot[{Exp[t/2]*Sin[t],Exp[t/2]*Cos[t]},
         {t,-10,1}]
```

Plot   The most basic plotting command.
```
     Plot[Sin[x], {x, -Pi, Pi}]
```

Plot3D   Creates 3-dimensional plot of a function of two variables.
```
     Plot3D[Sin[x]*Cos[y], {x, -Pi, Pi}, {y, -Pi, Pi}]
```

PlotLabel   Puts a label on a plot.
```
     Plot[{Sin[x], Cos[x]}, {x, -Pi, Pi},
         PlotLabel → "Trig functions"]
```

PlotRange   Controls how much of a plot to show. If only one range is given, it is applied
                to the vertical axis.
```
     Plot[AiryBi[x], {x, -10, 10}, PlotRange → {-5, 5}]
```

PlotStyle   Controls the style of a plot: color, thickness, whether solid or dotted, *etc.*
```
     Plot[{Sin[x], Cos[x]}, {x, -Pi, Pi},
         PlotStyle → {{Red, Dashed}, {Blue, Dotted}}]
```

Show   Combines plots or redisplays them with new options.
```
     Show[Table[pp[n, n/3], {n, 1, 3}]]
```

Ticks   Controls the tick marks along the axes.
```
     Plot[Sin[x], {x, -Pi, Pi}, Ticks →
         {{-Pi, -Pi/2, 0, Pi/2, Pi}, {-1, 0, 1}}]
```

VectorFieldPlot Plots vector fields. With the option **ScaleFunction**, you can
                rescale the arrows. In the **VectorFieldPlots** package.
```
     Needs["VectorFieldPlots`"]
     VectorFieldPlot[{x*(1 - x)*y, y^2}, {x, -1, 1},
         {y, -1, 1}, ScaleFunction → (1 &)]
```

Built-in Constants

E $e = \exp(1)$. You can also use the symbol **e** from the palette.

I $i = \sqrt{-1}$. You can also use the symbol **i** from the palette.

Infinity ∞. You can also use the symbol from the palette.

Pi π. You can also use the symbol from the palette.

Mathematica Programming

Do   Repeats a calculation a specified number of times.
```
Do[Print[i, " ", i^2], {i, 1, 10}]
```

For   Executes the first argument, then repeats a calculation (the fourth argument), increasing a counter according to the third argument, until a certain test (the second argument) fails to give **True**.
```
sum = 0; For[n = 1, 1/n^2 > 10^-3, n++, sum =
    sum + 1/n^2;]; sum
```

If   Conditional statement. When the first argument is **True**, the second argument is executed; when it's false, the third argument is executed.
```
sign[x_] := If[x < 0, -1, If[x == 0, 0, 1]]
Map[sign, Table[n, {n, -5, 5}]]
```

Message   Prints a warning message.
```
mysqrt::neg = "trying to take square root of a negative number";
mysqrt[x_] := If[x < 0, Message[mysqrt::neg], Sqrt[x]]
mysqrt[-1]
```

Module   Sets up a block of code for a subroutine, within which one can set up local variables.
```
PascalTriangle[n_] :=
Module[{array, i, j}, array = ConstantArray[0, {n+1, n+1}];
    array[[1, 1]] = 1; Do[array[[i, j]] =
      array[[i-1, j]] + If[j > 1,
      array[[i-1, j-1]], 0], {i,2,n+1}, {j,1,n+1}];
    array // MatrixForm]
PascalTriangle[10]
```

While   Repeats a calculation until a certain test fails to give **True**.
```
a = 1; b = 2;
While[Abs[b-a] > 10^-15, b = a; a = (a + 2/a)/2;];
N[a, 15]
```

Mathematica Packages

EquationTrekker Interactive package for studying ODE trajectories.

InterpolatingFunctionAnatomy Extra routines for understanding the inner workings of **NDSolve**.

`<<DifferentialEquations`InterpolatingFunctionAnatomy``

NDSolveUtilities Extra utilities for **NDSolve**, for example, for comparing results of different numerical methods.

`<<DifferentialEquations`NDSolveUtilities``

PlotLegends Package for adding legends to plots.

VectorFieldPlots Package for plotting vector fields.

Sample Solutions

Here we present complete solutions to selected problems from the Problem Sets. We prepared these solutions in *Mathematica* Notebooks and then reformatted them for printing.

Problem Set B, Problem 5

We are interested in the solutions to $y' = (t - e^{-t})/(y + e^y)$, particularly the solution satisfying $y(1.5) = 0.5$.

(a)

```
Clear[y]
sol5 = DSolve[y'[t] == (t - Exp[-t])/(y[t] + Exp[y[t]]), y[t], t]
```

Solve::tdep:

> The equations appear to involve the variables to be solved for in an essentially non-algebraic way. ≫

$$\left\{\left\{y[t] \rightarrow \text{InverseFunction}\left[e^{\#1} + \frac{\#1^2}{2} \&\right]\left[e^{-t} + \frac{t^2}{2} + C[1]\right]\right\}\right\}$$

The solutions of the equation are implicit, and the warning message indicates that *Mathematica* cannot find an explicit solution. We extract the implicit solution exactly as in Chapter 5, Example 5.5.

```
eqn5 = Solve[y == y[t] /. First[sol5], C[1]]
```

$$\left\{\left\{C[1] \rightarrow \frac{1}{2}\left(-2\ e^{-t} - t^2 + 2\left(e^y + \frac{y^2}{2}\right)\right)\right\}\right\}$$

So we define a function $f(t, y)$ equal to the right-hand side of the transformation rule reported by *Mathematica*. The solution of the IVP is obtained by setting this function equal to $f(1.5, 0.5)$.

```
Clear[f]
f[t_,y_] = C[1] /. First[impfunc] // Simplify
```

$$-e^{-t} + e^{y} + \frac{1}{2} \ (-t^2 + y^2)$$

So the general solution is $f(t, y) = c$, where f is as given.

(b)

We can use **ContourPlot** to plot some of the solution curves.

```
ContourPlot[f[t, y], {t, -1, 3}, {y, -2, 2}, Contours → 30,
    ColorFunction → "GrayTones"]
```

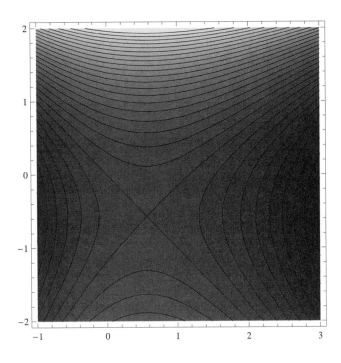

(c)

```
solcurve = ContourPlot[f[t, y] == f[1.5, 0.5],
    {t, -1, 3}, {y, -2, 2}, ContourStyle → Black]
```

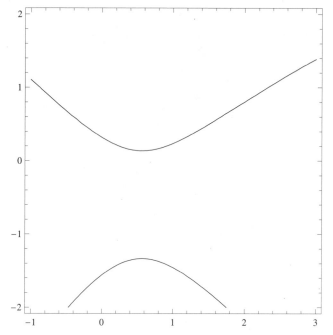

The graph shows the contour corresponding to the value $c = f(1.5, 0.5)$. The top curve corresponds to the solution of the IVP because it passes through the point $(1.5, 0.5)$.

(d)

```
tvals = {0, 1, 1.8, 2.1}
```

$\{0, 1, 1.8, 2.1\}$

Here are the values of the solution of the IVP at the points $t = 0, 1, 1.8,$ and 2.1.

```
yvals =
    Table[y /. FindRoot[f[tvals[[n]], y] == f[1.5, 0.5],
        {y, 0.5}], {n, 1, 4}]
```

$\{0.318386, 0.235633, 0.682187, 0.866212\}$

```
Show[solcurve,
    ListPlot[Table[{tvals[[n]], yvals[[n]]}, {n, 1, 4}],
        PlotStyle → {Black, PointSize[0.03]}]]
```

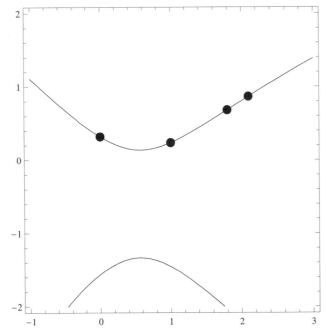

We have marked the points on the solution curve with t coordinates $0, 1, 1.8$, and 2.1.

Problem Set C, Problem 3

We are going to investigate the IVP $y' = (y - t)(1 - y^3), y(0) = b$, for various choices of the initial value b. We first try to find the exact (symbolic) solution with **DSolve**.

```
Clear[t,y]
DSolve[{y'[t] == (y[t] - t)*(1 - y[t]^3), y[0] ==b}, y[t], t]
```

$$\text{DSolve}\Big[\{y'[t] == (-t + y[t])\ (1 - y[t]^3), y[0] ==b\}, y[t], t\Big]$$

We see that **DSolve** is not able to find the exact solution. We thus solve the problem numerically.

In the next cell the first command creates a function (with delayed evaluation) representing the solution of the IVP for a given value of the initial condition $y(0) = b$, and the second command makes a list of solutions for $b = 0, 0.5, 1, 1.5$, and 2. We create this list so that *Mathematica* does not have to run **NDSolve** every time we use the function $y(t, b)$.

```
ya[t_,b_] :=
    y[t] /.
        First[NDSolve[{y'[t] == (y[t] - t)*(1 - y[t]^3),
            y[0] ==b}, y[t], {t,0,30}]]
sollist = Table[ya[t,b], {b, 0, 2, 0.5}]
```

NDSolve::ndsz: At t == 1.18797, step

    size is effectively zero; singularity or stiff system suspected. ≫

NDSolve::ndsz: At t == 2.11968, step

    size is effectively zero; singularity or stiff system suspected. ≫

```
{InterpolatingFunction[{{0.,1.18797}}, <>][t],
    InterpolatingFunction[{{0.,2.11968}}, <>][t],
    InterpolatingFunction[{{0.,30.}}, <>][t],
    InterpolatingFunction[{{0.,30.}}, <>][t],
    InterpolatingFunction[{{0.,30.}}, <>][t]}
```

Note that with the first two solutions, **NDSolve** stops before it reaches $t = 30$. As we shall see from the graphs, this is because these solutions have singularities near the indicated values.

(a)

Here is a plot of the five numerical solutions. The lowest curve corresponds to $b = 0$ and the highest to $b = 2$.

```
plot1 = Plot[sollist, {t, 0, 4}, PlotRange → {-3, 4},
    PlotStyle → Black]
```

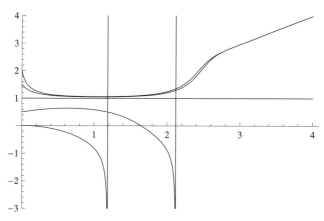

(b)

The plot indicates that if $b = 1$, the corresponding solution is $y = 1$; if $0 < b < 1$, the solution first increases and then decreases to $-\infty$; if $b \leq 0$, the solution decreases to $-\infty$; and if $b > 1$, the solution first decreases and then increases to $+\infty$. That $y = 1$ is the solution corresponding to $b = 1$ can be verified by substituting $y = 1$ into the IVP. It also appears that for values of b less than 1, the solutions have vertical asymptotes at finite values of t. One cannot be certain about this from the plots, but it is in fact true.

(c)

Now we'll combine our graphs with the graph of the line $y = t$.

```
plot2 = Plot[t, {t, 0, 4}, PlotStyle → {Black, Dashed}];
Show[plot1, plot2]
```

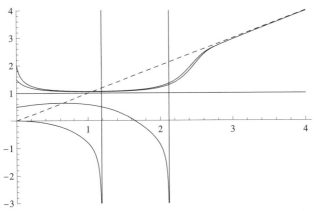

It appears that for $b > 1$, the solutions are asymptotic to the line $y = t$. This is further clarified by plotting the three larger solutions over a longer interval.

```
Plot[sollist[[{3, 4, 5}]], {t, 0, 30}, PlotStyle → Black]
```

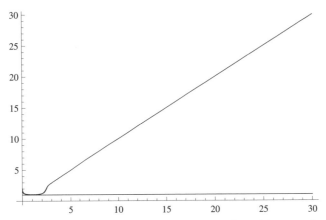

To understand this asymptotic behavior, note that y' is zero along the line $y = t$. When $y > 1$, y' is negative above the line $y = t$, and positive below it. Furthermore, as y gets larger, $|y'| > 1$ except when y is very close to t. Thus, the solutions are pushed toward the line $y = t$.

Here is a plot of the direction field for the differential equation, which confirms the observations we have made about the solutions. Notice the horizontal arrow at the upper right grid point; this occurs because the grid point $(4, 4)$ is exactly on the line $y = t$.

```
<<VectorFieldPlots`
VectorFieldPlot[{1, (y-t)*(1-y^3)}, {t,0,4}, {y,-1,4},
    ScaleFunction → (1&), PlotPoints → 20, Frame → True]
```

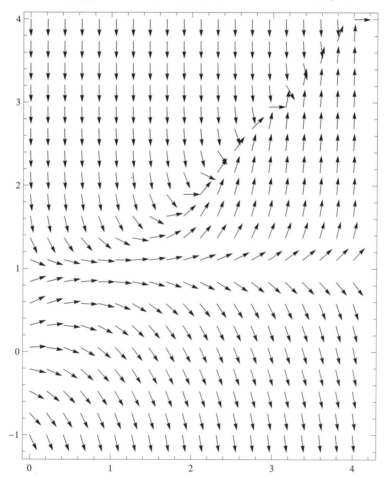

Problem Set D, Problem 1

We consider Airy's equation

$$y'' = ty$$

with initial conditions $y(0) = 0$, $y'(0) = 1$. We first find the solution.

```
y1[t_] =
    y[t] /.
        First[DSolve[{y''[t] == t*y[t], y[0] == 0, y'[0] == 1},
            y[t], t]]
```

$$\frac{1}{6}\left(-3\ 3^{1/3}\ \text{AiryAi}[t]\ \text{Gamma}\left[\frac{1}{3}\right] + 3^{5/6}\ \text{AiryBi}[t]\ \text{Gamma}\left[\frac{1}{3}\right]\right)$$

(a)

For t close to 0, we want to compare the solution of Airy's equation with the solution to
the IVP: $y'' = 0$, $y(0) = 0$, $y'(0) = 1$. The solution of this IVP is $y = t$. Here is a plot of
the solution to Airy's equation and the facsimile solution $y = t$. The facsimile solution is
plotted as a dashed curve.

```
Plot[{y1[t], t}, {t, -2, 2},
    PlotStyle → {Black, {Black, Dashed}}]
```

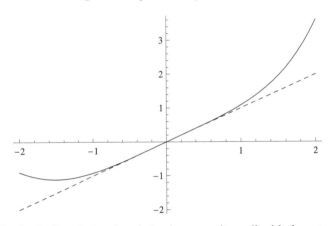

The facsimile solution does indeed agree quite well with the actual solution in a neighborhood of the origin.

(b)

Now for t close to $-16 = -(4^2)$, we want to compare the solution of Airy's equation to a
facsimile solution of the form $c_1 \sin(4t + c_2)$. We plot the solution of Airy's equation on
the interval $(-18, -14)$.

```
Plot[y1[t], {t, -18, -14}, PlotStyle → Black]
```

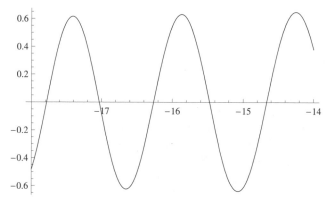

This certainly looks like a sine wave. Let's see how well it matches up with an appropriate sine wave. We have to choose constants c_1 and c_2 in the facsimile solution to make it match up. Note first that c_1 is the amplitude of the facsimile solution, and we can see from the graph that the amplitude of the solution to Airy's equation is about 0.625 on this interval (see the *Graphics* section of Chapter 8 for instruction on finding coordinates of points in *Mathematica* plots). The constant c_2 determines the phase shift and can be read off from the zeros of the solution. In particular, the solution of Airy's equation has a zero at about -16.25, so we should choose c_2 so that $4 * (-16.25) + c_2 = 0$; i.e. $c_2 = 65$. Here is the plot.

```
Plot[{y1[t], 0.625 Sin[4t + 65]}, {t, -18, -14},
    PlotStyle → {Black, {Black, Dashed}}]
```

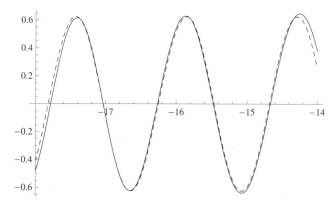

The plots match up very well. Why did we have to choose the values of c_1 and c_2 by hand? Our analysis suggests that the facsimile solution should approximate the actual solution near $t = -K^2$. But the initial condition we've used for Airy's equation is at the point $t = 0$, where the sinusoidal facsimile solutions do not approximate the solution of Airy's equation. Thus the undetermined constants in the facsimile solution do not really have anything to do with the initial conditions in Airy's equation, so we had to choose them by hand to make the solutions match up. Nevertheless, the *frequency* of the facsimile solution is determined by K and is independent of the undetermined constants. Thus we can conclude at least that the frequency of the solutions of Airy's equation in a neighborhood of $t = -K^2$ will be proportional to K.

(c)

Now for t close to $16 = (4^2)$, we want to compare the solution of Airy's equation to a facsimile solution of the form $c_1 \sinh(4t + c_2)$. As in (b), we first plot the numerical solution of Airy's equation on the interval $(14, 18)$.

```
Plot[y1[t], {t, 14, 18}, PlotStyle → Black]
```

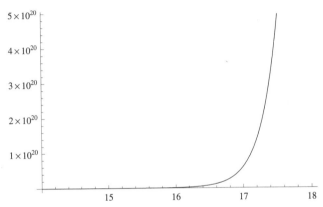

We'd like to compare this graph to that of the hyperbolic sine function. Again, we have to choose c_1 and c_2. Let's make an arbitrary choice this time: $c_1 = 1$, $c_2 = 0$.

```
Plot[Sinh[4t], {t, 14, 18}, PlotStyle → Black]
```

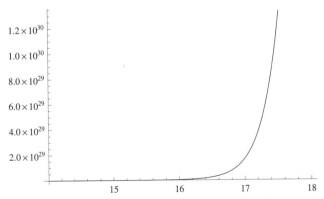

The graphs are remarkably similar. Note, however, that the values in the second graph are about $3 * 10^9$ greater than in the first, which means we could choose c_1 to be about $3 * 10^{-10}$ to make the values of the functions similar.

(d)

```
Plot[y1[t],{t,-20,2}, PlotStyle → Black]
```

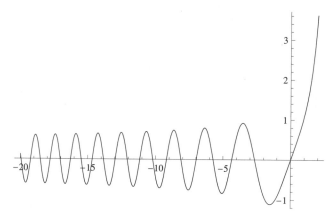

It appears from the graph that the frequency increases and the amplitude decreases as t goes to $-\infty$. The increasing frequency was predicted by our facsimile analysis. The decreasing amplitude is harder to explain.

Problem Set E, Problem 1

(a)

First let's define the left-hand side of the differential equation.

```
airy = y''[x] - x*y[x]
```

$-x\ y[x] + y''[x]$

Now we take the series expansion and substitute the initial conditions.

```
airyeqseries = Series[airy, {x, 0, 10}] /.
    {y[0] → 1, y'[0] → 0}
```

$y''[0] + (-1 + y^{(3)}[0])\ x + \dfrac{1}{2}\ y^{(4)}[0]\ x^2 +$

$\dfrac{1}{6}\ (-3\ y''[0] + y^{(5)}[0])\ x^3 + \dfrac{1}{24}\ (-4\ y^{(3)}[0] + y^{(6)}[0])\ x^4 +$

$\dfrac{1}{120}\ (-5\ y^{(4)}[0] + y^{(7)}[0])\ x^5 + \dfrac{1}{720}\ (-6\ y^{(5)}[0] + y^{(8)}[0])\ x^6 +$

$$\frac{(-7\ y^{(6)}[0] + y^{(9)}[0])\ x^7}{5040} + \frac{(-8\ y^{(7)}[0] + y^{(10)}[0])\ x^8}{40320} +$$

$$\frac{(-9\ y^{(8)}[0] + y^{(11)}[0])\ x^9}{362880} + \left(-\frac{y^{(9)}[0]}{362880} + \frac{y^{(12)}[0]}{3628800}\right)\ x^{10} + O[x]^{11}$$

Now solve for the coefficients.

```
coeffs = Solve[airyeqseries == 0]
```

$$\{\{y''[0] \to 0, y^{(3)}[0] \to 1, y^{(4)}[0] \to 0, y^{(5)}[0] \to 0,$$
$$y^{(6)}[0] \to 4, y^{(7)}[0] \to 0, y^{(8)}[0] \to 0, y^{(9)}[0] \to 28,$$
$$y^{(10)}[0] \to 0, y^{(11)}[0] \to 0, y^{(12)}[0] \to 280\}\}$$

Substitute the coefficients into the Taylor series of **y[x]** to get the Taylor series of the solution.

```
airyseries =
    Series[y[x] ,{x, 0, 10}] /. {y[0] → 1, y'[0] → 0} /.
      First[coeffs]
```

$$1 + \frac{x^3}{6} + \frac{x^6}{180} + \frac{x^9}{12960} + O[x]^{11}$$

Let's also write the Taylor approximation as a polynomial.

```
taylorpoly = Normal[airyseries]
```

$$1 + \frac{x^3}{6} + \frac{x^6}{180} + \frac{x^9}{12960}$$

(b)

First let's find the exact solution to the differential equation, using **DSolve**.

```
exactsol =
    y[x] /. First[DSolve[{airy == 0, y[0] == 1, y'[0] == 0},
      y[x] , x]]
```

$$\frac{1}{2}\left(3^{2/3}\ \text{AiryAi}[x]\ \text{Gamma}\left[\frac{2}{3}\right] + 3^{1/6}\ \text{AiryBi}[x]\ \text{Gamma}\left[\frac{2}{3}\right]\right)$$

Let's draw plots with the exact solution as a solid black line and the approximate solution as a dashed line.

```
Plot[{exactsol, taylorpoly}, {x, 0, 5},
    PlotStyle → {Black, {Black, Dashed}}]
```

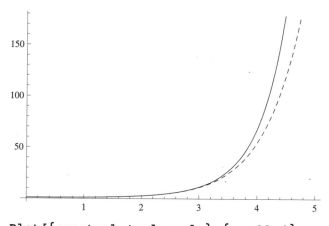

```
Plot[{exactsol, taylorpoly}, {x, -10, 0},
    PlotRange → {-3,1}],
    PlotStyle → {Black, {Black, Dashed}}]
```

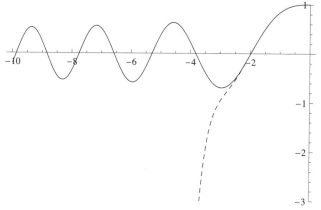

We see that the Taylor polynomial of degree 9 gives a good approximation to the correct solution over the interval $(-3, 3)$.

(c)

For $x > 1$, $y''(x) > y(x)$. Let's solve $y''(x) = y(x)$.

```
y[x] /. First[DSolve[y''[x] == y[x], y[x], x]]
```

$$e^x \, C[1] + e^{-x} \, C[2]$$

The second term dies rapidly, but the first term grows exponentially if $C[1] \neq 0$. In fact, we can compare the exact solution with an exponential by looking at a logarithmic plot. (If it were exactly a multiple of e^x, we'd see a straight line.)

```
LogPlot[{exactsol, eˣ, taylorpoly}, {x, 0, 5},
    PlotStyle → {Black, {Black, Dashed}, {Black, Dotted}}]
```

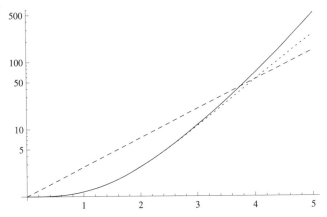

So the exact solution overtakes e^x around 3.7 and grows more rapidly after that. The Taylor series can't keep up with this rapid growth past this point.

(d)

As we saw in the graph above for negative x, the exact solution oscillates when $x < 0$. Since a polynomial of degree n can only have n roots, we need a Taylor polynomial of very high degree if we want to cross the x-axis many times. Furthermore, a polynomial must go to $\pm\infty$ as $x \to -\infty$, while the exact solution oscillates but approaches 0.

Problem Set F, Problem 6

(a)

First we find the critical points of the system.

```
Clear[x, y]
crit = Chop[Solve[{x(1.5 - x - 0.5y) == 0, y(2 - 1.5x - 0.5y) == 0},
    {x, y}]]
```

```
{{x → 0, y → 0}, {x → 0, y → 4.},
    {x → 1., y → 1.}, {x → 1.5, y → 0}}
```

Since we entered the coefficients as floating point numbers, *Mathematica* uses floating point arithmetic to solve the system, rather than exact arithmetic. This leads to round-off errors, so we use **Chop** to replace a spuriously small number (about 10^{-16}) in the output with 0. The critical points are $(0,0)$, $(0,4)$, $(1,1)$, and $(1.5,0)$. To analyze their stability, we compute the matrix of partial derivatives of the system, and then evaluate this matrix and find its eigenvalues at the critical points.

```
sys1 = x (1.5 - x - 0.5y) ;
sys2 = y (2 - 1.5x - 0.5y) ;
A = D[{sys1, sys2}, {{x, y}}]
```

$\{\{1.5 - 2\ x - 0.5\ y, -0.5\ x\}, \{-1.5\ y, 2 - 1.5\ x - 1.\ y\}\}$

Below we use the option **TableDepth** $\rightarrow$ **2** to prevent **TableForm** from removing all levels of braces.

```
TableForm[Table[{{x, y} /. crit[[n]],
        Eigenvalues[A /. crit[[n]]]}, {n, 1, 4}],
    TableDepth → 2, TableSpacing → {1, 5},
    TableHeadings → {None, {{x, y}, "eigenvalues"}}]
```

$\{x,y\}$	eigenvalues
$\{0,0\}$	$\{2., 1.5\}$
$\{0,4.\}$	$\{-2., -0.5\}$
$\{1., 1.\}$	$\{-1.65139, 0.151388\}$
$\{1.5, 0\}$	$\{-1.5, -0.25\}$

By inspecting the eigenvalues at the four critical points, we can classify them as follows:

$(0,0)$ unstable node
$(0,4)$ asymptotically stable node
$(1,1)$ unstable saddle point
$(1.5,0)$ asymptotically stable node

(b)

Here is a plot of the vector field of the system. We find that the vector field looks best when x and y have similar ranges, so we use a wider range of x than is needed to capture the critical points.

```
<<VectorFieldPlots`
csvf = VectorFieldPlot[
    {x (1.5 - x - 0.5y) , y (2 - 1.5x - 0.5y) }, {x, 0, 4},
    {y, 0, 5}, ScaleFunction → (# + 50&) , PlotPoints → {16, 20},
    Frame → True, FrameLabel → {x, y}]
```

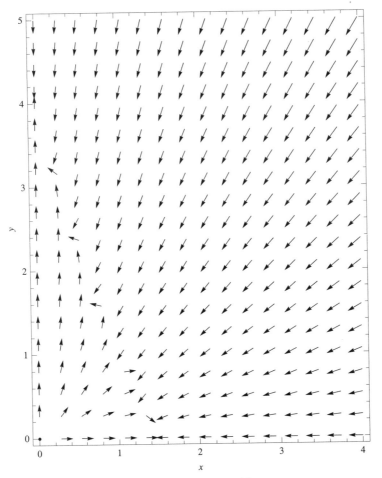

The vector field confirms what we found in part (a).

(c)

To plot the phase portrait of the system, we use the approach described at the end of Chapter 12. First we define a function that computes a numerical solution for the initial condition $(x0, y0)$.

```
Clear[t, x0, y0]
nsol[x0_, y0_] :=
    {x[t], y[t]} /.
        First[NDSolve[{x'[t] == x[t] (1.5 - x[t] - 0.5y[t]),
            y'[t] == y[t] (2 - 1.5x[t] - 0.5y[t]), x[0] == x0,
            y[0] == y0}, {x[t], y[t]}, {t, -5, 20}]];
```

Next we try a rectangular grid of 30 different initial conditions. The following commands generate a lot of warning messages due to solutions that go to infinity for negative t. We turn the warning messages off to reduce clutter (both here and in the remainder of the Notebook).

```
Off[NDSolve::ndsz, InterpolatingFunction::dmval]
trajs = Flatten[Table[nsol[i, j], {i, 0, 2, 0.5}, {j, 0, 5}],
    1];
ParametricPlot[Evaluate[trajs], {t, -5, 20},
PlotRange → {{0, 4}, {0, 5}}, PlotStyle → Black]
```

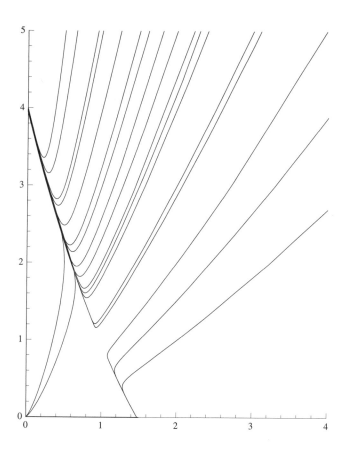

This graph is not ideal. In particular, we need more initial conditions near the origin. We can see from the graph that a better choice of initial conditions would be a collection of evenly spaced points along lines parallel to and above and below the line from $(0, 4)$ to $(1.5, 0)$. Since that line has slope about -2.5, we redefine the table of trajectories as follows:

```
trajs = Flatten[Table[nsol[j*x0, j*(2.5 - 2.5x0)],
    {x0, 0, 1, 0.1}, {j, 0.6, 2.6, 2}], 1];
csplot = ParametricPlot[Evaluate[trajs], {t, -5, 20},
    PlotRange → {{0, 4}, {0, 5}}, PlotStyle → Black]
```

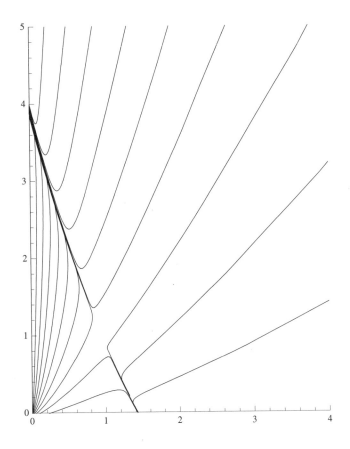

Much better! By putting the parametric plot and the vector field plot together we can see the direction of the trajectories.

```
Show[csplot, csvf]
```

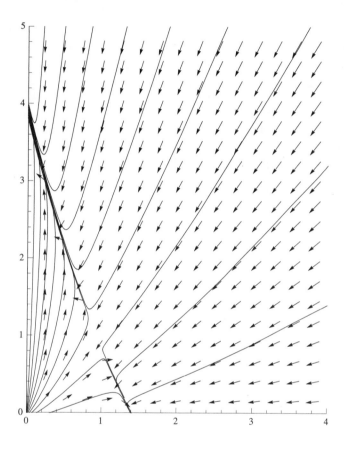

(d)

Here is a table of values of the populations.

```
TableForm[Table[Evaluate[{t, nsol[2.5,2]}], {t, 1, 20}],
    TableHeadings → {None, {"Time", "Populations"}},
    TableSpacing → {1, 5}, TableDepth → 2]
```

Time	Populations
1	{1.29442, 0.761067}
2	{1.20257, 0.629451}
3	{1.20352, 0.569922}
4	{1.2221, 0.520467}
5	{1.24499, 0.471819}
6	{1.2692, 0.422987}
7	{1.29376, 0.374566}
8	{1.31806, 0.32747}
9	{1.34158, 0.282626}
10	{1.36383, 0.240847}
11	{1.38442, 0.202758}
12	{1.40307, 0.168753}
13	{1.41961, 0.138989}
14	{1.434, 0.113408}
15	{1.4463, 0.0917775}
16	{1.45665, 0.07375}
17	{1.46524, 0.0589104}
18	{1.47228, 0.046823}
19	{1.478, 0.0370635}
20	{1.4826, 0.0292404}

(e)

There is no peaceful coexistence because almost all the trajectories tend toward an equilibrium point representing a positive population of one species and a zero population of the other. For example, in (d) we saw that the population of x tends to 1.5 and the population of y tends to 0. The only non-zero initial population distribution which results in no change is $x = 1, y = 1$.

To approximate the separatrices, we solve the system for negative time for initial conditions near the unstable equilibrium point $(1, 1)$. From the portrait in (c), we expect one separatrix running from the origin to $(1, 1)$; and another running from $(1, 1)$ out towards infinity. To find the first we choose an initial data point just to the left and below $(1, 1)$ and solve backwards in time; and to find the second we do the same thing with an initial data point just above and to the right of $(1, 1)$. We plot the separatrices as thick lines.

```
separatrixplot=
    ParametricPlot[
        Evaluate[{nsol[0.99, 0.99], nsol[1.01, 1.01]}],
```

```
      {t, -5, 0}, PlotRange → {{0, 4}, {0, 5}},
      PlotStyle → {{Black, Thick}}];
Show[csplot, csvf, separatrixplot]
```

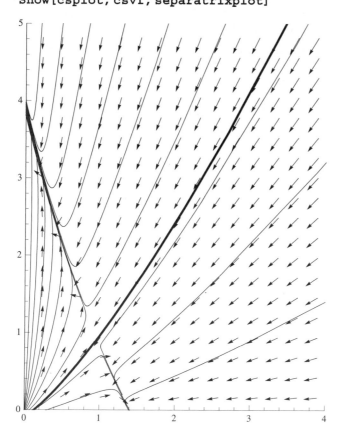

(f)

The following commands use event detection, as described in Section 8.12.1, to stop the numerical solution when x reaches 1.5. We start even closer to the fixed point $(1, 1)$ to get greater accuracy than we used for the figure above.

```
nsolivp =
    First[NDSolve[{x'[t] == x[t] (1.5 - x[t] - 0.5y[t]),
        y'[t] == y[t] (2 - 1.5x[t] - 0.5y[t]), x[0] == 1.0001,
        y[0] == 1.0001}, {x, y}, {t, -10, 0},
        Method → {"EventLocator", "Event" → (x[t] - 1.5)}]]
```

```
{x → InterpolatingFunction[{{-4.95197, 0.}}, <>],
   y → InterpolatingFunction[{{-4.95197, 0.}}, <>]}
```

Notice that the numerical solution is defined only to about $t = -4.95$, suggesting that this is (approximately) the time at which $x(t) = 1.5$. We can compute this time, and the corresponding value of y, more exactly as follows.

```
<<DifferentialEquations`InterpolatingFunctionAnatomy`

stopt =
    First[First[InterpolatingFunctionDomain[x /. nsolivp]]]
{x[stopt], y[stopt]} /. nsolivp
```


```
-4.95197
```

```
{1.5, 1.69338}
```

Indeed, our approximate separatrix crosses the line $x = 1.5$ at about $y = 1.69$. We could find the crossing point with greater accuracy by starting the solution even closer to the fixed point and making the error tolerances smaller in **NDSolve**, but with numerical methods we can't be absolutely sure how close our approximation is to the true crossing point.

Index